HOW TO REPAIR
HERSCHEDE
TUBULAR BELL CLOCKS

EXPANDED SECOND EDITION

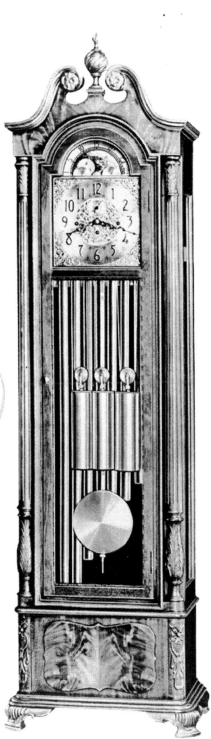

STEVEN G. CONOVER

Books by Steven G. Conover

Repairing French Pendulum Clocks $24.95
How to Repair 20 American Clocks $29.95
Clock Repair Basics $22.95
Clock Repair Skills $24.95
Striking Clock Repair Guide $22.95
Chime Clock Repair $28.50
How to Repair Herschede Tubular Bell Clocks $28.50
Building an American Clock Movement $21.95
How to Make a Foliot Clock $22.95

Clockmakers Newsletter Workshop Series:
Book 1 Repairs $37.95
Book 2 Tools, Tips & Projects $37.95
Book 3 Escapements $34.95
Book 4 Grandfather Clocks $34.95
Book 5 Tubular Bell Clocks $32.95

New books in the Workshop Series will be announced as they are published.

Steven G. Conover's books are available through our web store: www.clockmakersnewsletter.com.

The books are also available by mail order from:
Clockmakers Newsletter
203 John Glenn Avenue
Reading PA 19607
(610) 796-0969

And from clock parts suppliers

HOW TO REPAIR
HERSCHEDE
TUBULAR BELL CLOCKS

SECOND EDITION

STEVEN G. CONOVER

CLOCKMAKERS NEWSLETTER, READING

Dedicated to my wife, Karen, for being the best proofreader and advisor.

Published by Clockmakers Newsletter
203 John Glenn Ave.
Reading PA 19607

Second Edition, 1999
Text and Illustrations Copyright 1986, 1989, 1999 by Steven G. Conover

ISBN 978-0-9624766-7-9

CONTENTS

INTRODUCTION to the SECOND EDITION

Herschede grandfather clocks have been with us since 1885. The company started out by designing the cases, which were made locally in Cincinnati, and installing imported movements. In 1911, Herschede began installing its own movements. After many years of production, the company moved from Cincinnati to Starkville, Mississippi in 1960. Herschede continued to build its famous tubular bell clocks at the new factory until 1984. New owners restarted production a few years later, but the effort was short-lived. The present Herschede owner supplies parts for Herschede movements.

This book is about the Herschede tubular bell movement—how to clean, repair, assemble, and test it. We will not concern ourselves with halfway or "first-aid" repairs intended to postpone an overhaul. When it is time to clean the movement, nothing else will do. To clean the movement, you must take it completely apart. Even if you use an ultrasonic cleaning machine, disassembly is still just as important. When the movement is apart, you can take care of any repairs which may be necessary. This book, then, takes you step-by-step through an overhaul of a Herschede tubular bell movement. Assembly and adjustment procedures are covered in detail.

This second edition includes the basic repair information from the 1986 edition, but much additional material has been added. A chapter on the two-weight tubular bell movements will answer the requests of many repairers. The troubleshooting chapter has been expanded. New information on refinishing the plates and wheels will help those who need to completely restore a damaged movement. Other all-new material is a reproduction of the Owner's Manual, a company history, plans for a factory test stand, and a listing of serial numbers. Added to the movement drawings are images and part numbers for individual parts, to help in the ordering of replacements from the present-day Herschede operation. Finally, a chapter is devoted to pictures and catalog descriptions of many of the Herschede tubular bell clocks from the last decades of the company's output.

What experience do you need before repairing a Herschede tubular bell clock? Well, one look at one of these large and imposing hall clocks should be enough to convince you that it is not a task for a beginner. If you have disassembled and overhauled a number of chime clocks, you may be ready. It is impossible to set a minimum level of experience—one year, three years, or any other period. The quality of the repairs you have done is the best yardstick. Some repairers learn very little as they go along, take shortcuts as they find them, and are in for trouble. Others take each job seriously, and do not hesitate to tackle any job for which they are qualified. So there it is. You must judge your own readiness.

Just what is it that makes a Herschede tubular bell clock different from other chime clocks? For one thing, the clock itself is more expensive than most other clocks you will ever work on. In the last years of production, most other makers' new grandfather clocks cost less than the Herschede movement alone. Your customer knows the clock is valuable, and he or she will certainly insist that it work perfectly after you have finished repairing it. The same holds true for any grandfather clock and its owner, but a service problem with a Herschede always seems more acute. People expect the very best from it. Another feature of the Herschede is the extra time and effort required for take-down and packing, then re-delivery and set-up. The movement, weights, and tubes are especially heavy and bulky to handle. Repeat service calls can be very inconvenient, to say the least.

I hope you will find this book to be a complete and invaluable reference guide to the repair of the Herschede tubular bell clock.

Steven G. Conover
Reading, Pennsylvania, August 1999

1

HOW TO GET STARTED

Fig. 1. Detail of 5-tube clock dial, 1920's

Let's assume you have agreed to repair a Herschede grandfather clock. You go to the customer's home to check the clock over thoroughly. The complaint, by the way, is that the clock chimes and strikes slowly, and frequently stops for no apparent reason. For the sake of our discussion, we'll work on a 9-tubular bell model. At this beginning stage of the repair job, it is essentially the same as a 5-tube clock, except for the chore of handling more tubes. So, you remove the hands and dial, and you are confronted by the movement which looks exactly like the one in Figures 2 and 3. What do you do first?

To begin, you should see whether the movement is simply dry of oil. Oiling will cure the clock's prob-

lems only if the movement is still reasonably clean and seems to need the oil. A dirty, gummy movement should not be oiled again. You must do the complete cleaning job. A clock this valuable is worth an overhaul! Another reason to recommend an overhaul would be any obvious and serious damage you can detect, and could not possibly repair in the customer's home. Bent or broken pivots, damaged gears, missing parts, or signs of severe abuse are examples.

Let's say you have determined the clock movement is dirty, and it shows some wear. It needs an overhaul. The customer agrees with you, and approves the job, including the price. You carefully remove the movement from the clock. You pack the

Fig. 2. Front view, 9-tube Herschede movement, 1980's.

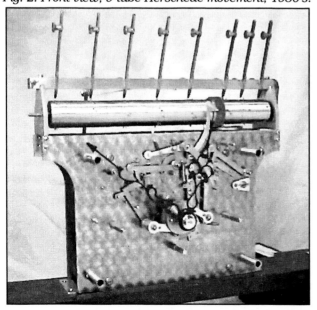

Fig. 3. Rear view of the movement

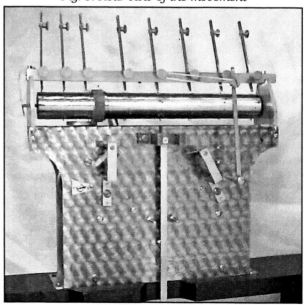

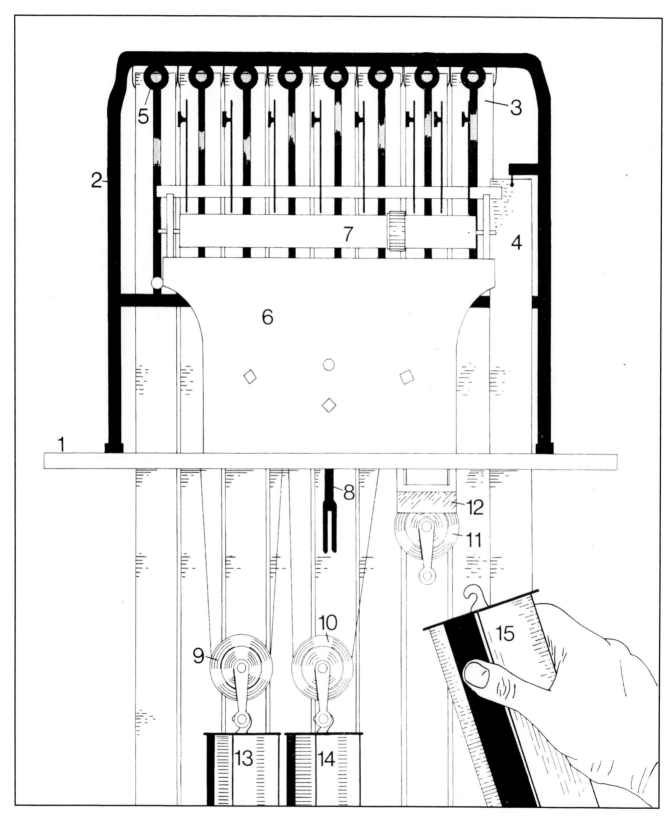

Fig. 4. Taking the movement out of the cabinet (see text)

1 seatboard	4 hour tube	7 cylinder	10 time pulley	13 strike train weight (marked S.T.)
2 tube rack	5 chime hammer	8 fork	11 chime pulley	14 time train weight (marked T.T.)
3 chime tube	6 movement	9 strike pulley	12 wooden pulley block	15 chime train weight (marked C.T.)

movement and seatboard (with tube rack attached), weights, pendulum, dial, and hands, to take them to your shop.

This may seem easy, but it is not. You will want to remove the "works" as neatly as possible. Do it as though you have done it many times before, even if you haven't. Don't rush. Take deliberate steps and be careful. Don't give the customer the impression you are about to lift the engine out of an old car, when you have no idea what to unhook first. There are several ground rules. Don't bang the tubes together—this unnerves the customer. Handle the weights with soft cotton gloves. And pack everything well to prevent damage.

The general "takedown" procedure goes as follows. Refer to Figure 4. As mentioned above, remove the hands and dial first. To remove the hands, take out the tapered pin, then the minute hand. Pull off the hour hand and the second hand. The dial is released as you unhook the four dial post knobs, one behind each corner of the dial. (If you have trouble removing the dial, consult Chapter 8.) Install the Herschede pulley blocks (12) on each of the pulleys (9, 10, 11), then wind the cables up until the blocks rest up against the underside of the seatboard (1).

The wooden pulley block (Figure 5) is a wonderful idea. One of its functions is to prevent the weight cable from becoming tangled behind the main wheel. On the chime train, it can be especially difficult to free the cable, which can become wedged between the second wheel and the front of the cable drum. In addition, the block stops the brass pulley from swinging about as you lift the movement. It's part of a professional job to use a set of three pulley blocks each time you remove or install a Herschede movement. Figure 6 gives the dimensions of the pulley block.

Referring back to Figure 4, now remove the three weights (13, 14, 15), bearing in mind that they are heavy and must be handled firmly. The chime weight, in particular, must be handled deliberately because it is so heavy. Wrap the three weights in clean paper. Ideally, you should box them, and tape the boxes. Do this in the customer's home. There must be no doubt in the customer's mind that you are taking good care of everything. Take the pendulum out of the clock. Remove the suspension spring and wrap the bob. Remove all of the tubes you can get out through the front of the cabinet. Perhaps you will have only the hour tube and the two longest chime tubes left in the clock. Your tube removal procedure will vary because different Herschede cabinets have their own clearances. Lay the tubes down carefully, safely out of the way.

The next part requires special care. Before you

Fig. 5. Wooden pulley blocks: original Herschede block (left) and homemade equivalent (right).

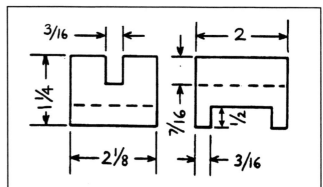

Fig. 6. Homemade wooden pulley block for use on Herschede tubular clocks. Dimensions are in inches.

go on, check under the clock to see whether it has the sharp spike feet which used to be popular for Herschede clocks placed on carpeted floors. Figure 7 shows one of these spikes. Imagine the damage it could do to just about any kind of floor! Blunted feet are more commonly seen. They still give good results as far as leveling is concerned and will not mar the floor. If you find the spike feet, proceed with extra caution or remove them before you move the

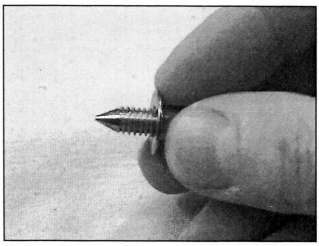

Fig. 7. Before moving the clock, check to see whether the clock has spike feet like this one!

clock around. (Once I was called in to check on a clock installed by someone else, and found the clock standing on spike feet—on a parquet floor!) Now, remembering that you have perhaps three tubes left in the clock, tilt the cabinet and pivot it on one front corner, moving it away from the wall.

Your object is to get at the back of the cabinet. Take off the four wood screws and remove the back panel. Carefully take out the remaining tubes through the back of the clock. Be careful not to bump them into the ceiling.

Carry on by removing the four seatboard screws and easing the movement (6), seatboard (1), and tube rack (2) out of the back of the clock, as a unit. Be careful with the fork (8); it extends below the seatboard. The unit must be lifted slightly to allow the fork to clear the lower edge of the opening in the back of the cabinet. Put the movement in a box you have prepared in advance and placed nearby. Tie down the fork, strike hammer arm, and chime hammers (5). Pad everything well for the trip to your shop. The customer has probably watched your every move in this whole procedure. Don't let it bother you. Most people are just fascinated to find out what is inside the clock cabinet.

Tell the customer that you will need at least a month or more to get the clock back to him or her. Try not to let the customer tie you down to a specific date. You can't possibly know what delays may arise.

2

DISASSEMBLY & CLEANING

Taking the movement apart is not quite the first step in our shop overhaul. It might be well to devote a few minutes first to inspecting the movement. Look for any obvious damage you missed when you checked the movement in the cabinet. If you find it necessary to order a part, even if it is just a set of cables, you should try to identify it now, so you won't have to delay the whole job later, waiting for the part to arrive. Inspecting all incoming repairs would save time in the end. The problem seems to be that most of us would rather start the repair jobs than inspect all the clocks without making any visible progress on their repair.

DISASSEMBLY

Following the initial inspection of the movement, you are ready to take it apart. First, remove the hammer cords. Also remove the hands and dial, if you haven't done so already. Unhook the cables from the winding drums and take the movement off the seatboard. Put the dial, seatboard and tube rack in an out-of-the-way place where they will not be damaged. On the following two pages, Figure 8 (front view) and Figure 10 (rear view) show the major parts you will be removing as you proceed.

Remove the eight screws and then the cylinder and hammer assemblies as a unit. Next, remove the anchor bridge and the anchor assembly, which includes the fork. It is important to use the right screwdriver for the anchor bridge screws, so you will not mar the slots. Most screwdrivers have a tapered tip which will fit many different screws fairly well but do not fit any screws exactly. The Herschede brass screw slots will mar easily, so a better idea is to modify a screwdriver so the tip has parallel sides. Use an old screwdriver or buy an inexpensive one for the purpose. Grind or file the tip so that it per-

fectly fits the Herschede screw heads. The same screwdriver should fit the pillar screws as well. Be careful as you remove the anchor, because the maintaining power mechanism exerts a small amount of force on the time train. Its purpose is to keep the clock running during winding, when the turning of the key neutralizes the force of the weight. The two fan bridges and the fan shaft assemblies can come off now, followed by the strike hammer arm.

Next, remove the front movement parts, beginning with the two racks, rack hooks, silent levers, and double lever. Herschede tubular movements have wires bent into an "s" shape, instead of tapered pins, to hold the levers on their studs. Stay with this practice, but don't reuse the old wires. Make up new ones when you reassemble the clock. Be careful not to lose any washers. Don't be tempted to leave the levers on, trusting an ultrasonic cleaner to get all the dirt out. Cleaning is only half the job. You might find a rough spot or burr, or some corrosion to be polished out. It's worth the full treatment. The two gathering pallets should be removed now. Take off the motion work—the intermediate bridge and wheel, the hour tube assembly (with the snail), and the hour bridge. Remove the minute tube assembly.

Now turn the movement onto the front, always taking care with the center arbor, which can bend. Remove the rear pillar screws. There are four brass screws, one at each pillar, and a steel screw at the spreader post, which is a fifth pillar supporting the plates. Carefully lift off the back plate and set it aside. Now take out the wheels.

You should remove the triangle-shaped second shaft bridge assemblies. There is one of these on

(text continued on page 8)

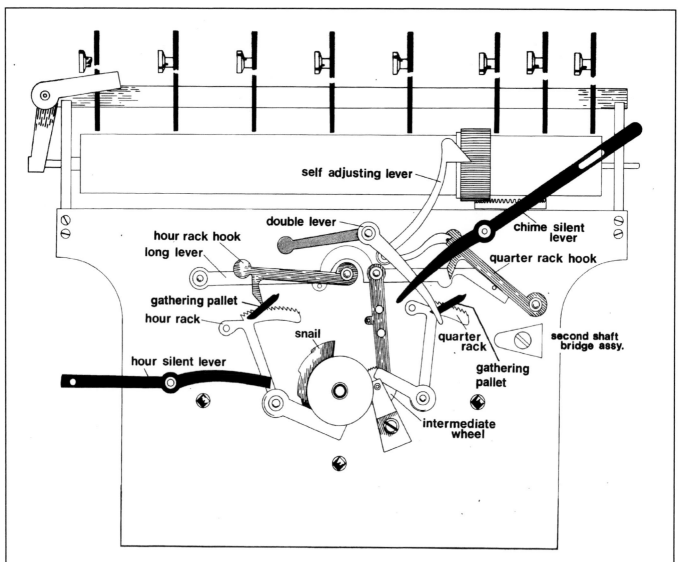

self adjusting lever

double lever

hour rack hook
long lever

chime silent
lever

quarter rack hook

gathering pallet

hour rack

snail

quarter
rack

second shaft
bridge assy.

gathering
pallet

hour silent lever

intermediate
wheel

Fig. 8. Front view of the 9-tube movement, showing the main parts for disassembly. As described in the text, parts should be taken from the outside and top of the movement before the plates are separated and the gears removed.

Fig. 9. The second shaft bridge assembly is also shown on Figure 8 above. The same part is found on the front and rear plates. Remove them to ensure complete cleaning and drying underneath.

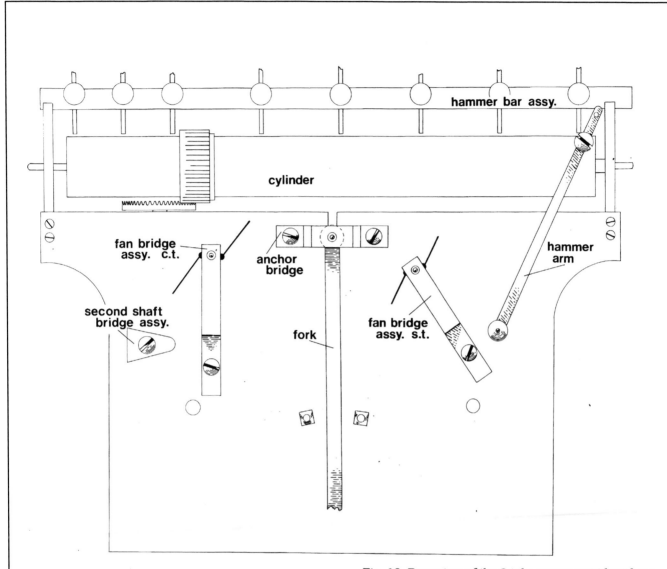

cylinder

hammer bar assy.

fan bridge
assy. c.t.

anchor
bridge

hammer
arm

second shaft
bridge assy.

fork

fan bridge
assy. s.t.

Fig. 10. Rear view of the 9-tube movement identifying the main parts to be removed before the plates are separated. See the text for a disassembly sequence for the main parts.

Fig. 11. One of the main wheel assemblies (the strike) is shown disassembled and ready for cleaning. The main wheel spring washer and main washer screw are shown in the photo. These are removed to permit the separation of the drums from the main wheels.

...text continued from page 5

the front plate and another on the rear plate. The parts are identified on Figures 8, 9, and 10. The two second shaft assemblies are the same part number, 2631. They fit over the chime second arbor pivots and permit fine-tuning the endshake of the arbor. Solvent will remain trapped under the second shaft assemblies unless you remove them before cleaning. In addition, they must be removed to permit the oiling of the chime second arbor pivots.

The three drums should be separated from the main wheels for the same reason, to prevent solvent from becoming trapped between the wheel and the drum. Separation also permits complete cleaning of deposits and later greasing of the parts for smooth action. The main shaft assemblies are made up of the drum, wheel, main shaft (winding arbor), click, and in the case of the time train main shaft assembly, the maintaining power mechanism. For each assembly, remove the main washer screw HER-2063, then push aside the main wheel spring washer HER-2062. The main parts can be separated now. Figure 11 shows the strike main shaft assembly taken down and ready for cleaning. The solvent will flow freely in and out of the drum through the notch for the cable, and should not pose a problem.

CLEANING

Cleaning is certainly a sensitive subject. Everyone seems to have a favorite method. In the early 1980's, Herschede published a recommended procedure which emphasized care of the finish. The manufacturer wanted the repairer to get the movement clean, but rightly expected him or her to worry about the lacquered brass parts at all times. Herschede's suggestion was to get the tough dirt off the movement by hand, using a brush and rinse solution, and then to peg out the pivot holes. They recommended cleaning the movement ultrasonically for a very short time. The procedure said: "one-half to three minutes, depending on newness of solution." The idea was to take off most of the dirt in advance, to minimize the ultrasonic cleaning. A short contact time with the solvent lessens the chance of damaging the finish on the plates.

All of this makes good sense, but you might prefer to rely a little more on the cleaning machine, if you have one, and less on hand brushing. Leave the parts in the solution for the shortest possible cleaning period. I am not going to quote ultrasonic cleaning times because the machines and solvents vary. Just remember to preserve the finish on the plates. One idea is to clean the plates by hand, brushing them for the shortest possible time, while cleaning most of the other parts ultrasonically.

What kind of solution should you use? There are two main types: ammoniated, which is water-based; and petroleum-based. Used according to directions, either type can be put in an ultrasonic cleaning machine. For hand brushing, I wouldn't want to soak my hands in either one for very long. Whatever your own cleaning procedure may be, you should use professional solvent and rinse solutions purchased from a dealer. Homemade solutions probably aren't worth the trouble and could damage the clock (or you).

Sometimes, careless repairers leave "high water" marks and green corrosion stains on Herschede movements. The plates are large, and you must take care not to let them stay half covered by the solvent. (Plates that have been ruined can be refinished as described in Appendix E.) The only metal part not to be immersed is the cylinder. Any solvent working its way into the cylinder might stay inside, or it might ooze out after causing corrosion. Just brush the cylinder clean, making sure that dirt and grease are removed from the pivots and the pins.

Drying the movement is important. A clock drying machine is one answer. You can make a movement dryer yourself out of a wooden box and an old style hair dryer, the kind with the hose connected to a cap (if any of these can still be found.). Remove the hat from the hose and insert the end of the hose into a hole in the top of the wooden box. A raised floor of metal screening keeps the parts from resting in pooled solutions that have dripped off the parts.

The next phase of the repair is to check the clean movement parts to see what repairs are needed. That's where the next chapter begins.

3

BASIC REPAIRS

Now your task is to find and correct any internal movement problems. If you miss the opportunity, all your good cleaning work and the adjustments which follow later may be wasted. One bad pivot can stop the clock.

PIVOTS

What can you look for? Check each and every pivot. Figure 12 illustrates a "bad" pivot. Run your fingernail over the pivot surface, looking closely and feeling for ridges, burrs, or any other imperfections. Each pivot must work easily in its hole. The higher up in the gear train, the more critical a smooth pivot becomes. This is not to say that any rough pivots in the lower arbors should be ignored. If the pivot roughness is manageable, polish it out on the lathe. But rather than cutting down the diameter of the pivot too much, consider repivoting or ordering an entire new wheel and arbor. Your choice will depend on cost and whether you can repivot accurately and with a reasonable effort. The Herschede is a modern movement, and parts should be available for some time to come.

Check each pivot by installing the arbors between the plates, one at a time. The arbor should turn freely, gliding to a gradual stop. The arbor should have endshake, which is front to back freedom. Any

tightness means you have rough spots on one or both pivots, or an ill-fitting bushing from a previous repair. It may be that the pivot is good, but the hole is worn.

BUSHINGS

It is sometimes necessary to install replacement bushings in a Herschede movement. If you have found a worn hole, double check the pivot for a good finish. It is important to start with a good pivot, for a rough one will rapidly cut into the new bushing. Next, you have to decide how you will install the bushing.

Press-Fit Bushings

The difference in bushing a Herschede tubular bell movement, compared to others, is the thickness of the plate. It is almost 4 mm thick. This puts it out of the range of most press-in bushings. The best solution is to use bushings made for Herschede movements. Some years ago, Herschede made bushings with outer diameters listed as "A" for the third

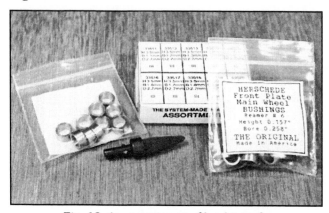

Fig. 13. An assortment of bushings for
Herschede tubular bell movements

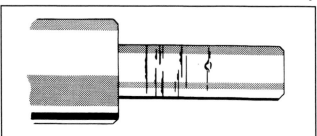

Fig. 12. A damaged pivot

wheel and up, through the largest, "D", for the winding arbor pivots. The bores of the "A" bushings were furnished very small, and they had to be reamed out to fit the pivot in each case.

Today, bushing assortments are available for Herschede (Figure 13). The bores are various sizes, and less reaming and finishing are required.

After you install a bushing, always fit the arbor between the plates by itself. Check for endshake. Spin the wheel, and notice whether it seems free. Turn the plate with the front up, then down, as the wheel spins, to observe any tendency to bind in a particular position. The bushing should be flush and smooth with the inside surface of the plate. Make sure the bushing allows the end of the pivot to extend outward at least a small amount. A bushing that is too long will allow the pivot to tunnel in, eliminating endshake on the arbor and causing a movement to fail.

Custom Made Bushings

It is occasionally necessary to make one or more bushings on the lathe. The following example is taken from a Herschede tubular bell movement whose plates had been rebushed with ten large, ugly screw-in bushings (Figure 14 shows one of

Fig. 14. One of ten screw-in bushings removed (left); a replacement bushing (right)

them). There was even some pivot damage resulting from the installation of these bushings. The bushings needed to be replaced because of their botch-work appearance and poor fit. I removed them, but the holes in the plates, besides being threaded, were much too large for press-fit bushings. Figure 15 shows the large hole left by a bushing that was unscrewed from the plate, as well as a custom made replacement bushing.

The threaded bushings were .193" in diameter. This meant the replacements had to be large diameter, thick-walled bushings. I decided to drill out the threaded bushing holes with a #10 drill (.1935").

It is important to pause before drilling and consider preparing a drill for brass. Twist drills can "grab" in brass and may tear the piece loose from the drill press and injure the operator. Figure 16 (top) shows a drill being rested flat on the bench. Rotate the drill so that one cutting edge, or lip, is parallel to the bench. Make a few light strokes with a sharpening stone across the lip. Hold the stone parallel to the bench as you bring it across the steel. Rotate the drill to bring the other lip into view, and repeat the procedure. The drill is now much safer to use on brass, but it is no longer good for drilling

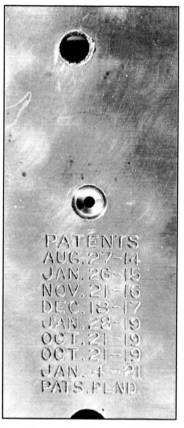

Fig. 15. A portion of the Herschede back plate, showing a threaded bushing hole before drilling and reaming (top) and a finished replacement bushing made on the lathe (center).

PATENTS
AUG. 27-14
JAN. 26-15
NOV. 21-16
DEC. 18-17
JAN. 28-19
OCT. 21-19
OCT. 21-19
JAN. 4-21
PATS. PEND.

steel. The modified drill is shown in Figure 16 (bottom).

I had already wondered whether any of the screwed bushings might have been installed off-center, but I decided it was better at this point to drill the new holes over the centers of the threaded ones and install new bushings. If a gear depthing problem became apparent after the wheels were tried in the movement, it would still be possible to remove a

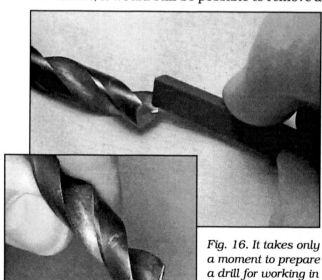

Fig. 16. It takes only a moment to prepare a drill for working in brass. This drill is now less likely to "grab" in the brass.

bushing and correct the depth (center distance between wheel and pinion). An undrilled bushing could be pressed in and drilled after the correct hole location was established and marked with a depthing tool. As it turned out, I did not discover any depth problems later on.

To prepare the Herschede plates for drilling, I removed the movement and dial pillars from the front plate. The plate could then be drilled from the

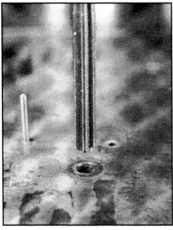

front surface without interference from riveted posts. A steel rod was installed in the drill press chuck as a locator. Once the plate was positioned, I clamped the plate and drilled the hole with the modified #10 drill. This was followed by the 13/64" reamer shown in Figure 17. A reamer can be expected to open the hole

Fig. 17. Reaming the plate

very close to what we want, but the task could have been done with a drill. All the holes that were to be rebushed were drilled and reamed in this manner.

Next, a length of 1/4" (.250") brass rod was mounted in the 3-jaw lathe chuck and turned down to between .204" and .205" diameter. This range of diameters could be pressed into the clock plate. At less than .204", the bushing would be loose.

A twist drill two sizes smaller than the finish pivot hole drill was used to drill through the bushing, followed by the finish drill. Whenever possible, I made two bushings with the same bore at once, by drilling deep enough to allow two bushings to be parted off from the brass rod. This job required six finish drill sizes, in the range of #47 to #60, for ten bushings. These small drills did not need to be modified for drilling brass, as the possibility of grabbing during the drilling operation is much less with small drills. Each bushing was parted off to a length of .142" to match the clock plates.

All the bushings were pressed into the plates, and then the oil sinks were cut with a bushing tool fitted with an oil sink cutter (Figure 18). A finished bushing (Figure 15) was hard to distinguish from an original pivot hole, despite the relatively large O.D. that was made necessary by the

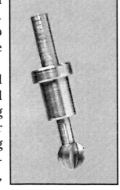

Fig. 18. An oil sink cutter

threaded bushings that were replaced.

Finally, each pivot was carefully fitted to its bushed hole. Several holes needed to be opened slightly with a hand-held cutting broach.

THE ANCHOR

Always check the pallet faces on the anchor, part #HER-2086, shown in Figure 19. If the pallets are grooved as shown in Figure 20, order a new anchor assembly from the factory. The Herschede escape-

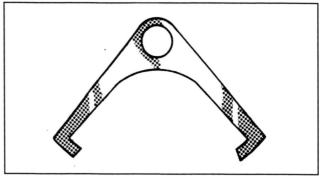

Fig. 19. Herschede solid deadbeat pallets, #HER-2086

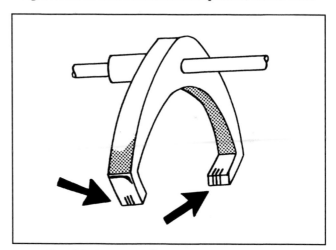

Fig. 20. A worn pallet unit must be replaced.

ment is the deadbeat type. This means that you cannot simply polish grooves out of the pallet faces and expect the escapement to work properly. A tubular bell movement which is in good condition otherwise can fail because of grooved pallets.

It is possible to replace the anchor in the customer's home. It isn't the easiest job. You have to obtain access to the back of the clock, and after you have removed the weights, you must remove the anchor bridge. The anchor can be taken out at this point, but be aware of a small amount of power contained in the maintaining power assembly.

Anchor Assembly HER-22612

There is another type of pallet unit that was made for a relatively small number of Herschede tubular

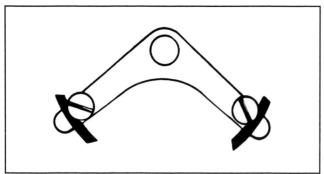

Figure 21. The pallet unit HER-22612 cannot be used in place of the pallet unit shown in Figure 19 (see text).

bell movements. Called the Vulliamy type, after its inventor, Herschede's version of this pallet unit is part HER-22612 (Figure 21). It has adjustable, replaceable pallet pads screwed to a brass pallet body. It is important to note that this pallet unit is not interchangeable with the solid anchor HER-2086 (Figure 19). A different escape wheel, anchor bridge, and pendulum assembly were used with each type of anchor. Herschede used anchor HER-22612 in the 1970's in some tubular movements with a self-setting beat escapement. The center distance of the escapement will be wrong if you install an anchor HER-22612 in place of a solid one, HER-2086.

Adjusting the Anchor

Returning to the solid anchor HER-2086, there is only one adjustment which can be made to this deadbeat escapement. It's worth reassembling the time train from the third wheel upward and checking it now. The center distance of the pallet unit to

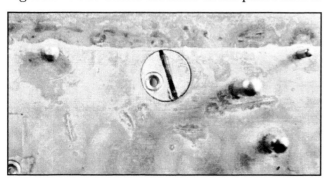

Fig. 22. This eccentric bushing slot was damaged by a screwdriver blade that did not fit.

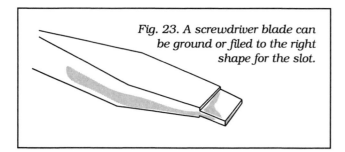

Fig. 23. A screwdriver blade can be ground or filed to the right shape for the slot.

the escape wheel can be changed. Observe the escapement, and if you see the escape wheel teeth drop further onto one pallet than the other, you can adjust the depth to equalize this drop. The adjustment is made by turning an eccentric bushing for the front of the pallet arbor.

Figure 22 shows the upper front movement plate area on a movement which was improperly adjusted. The eccentric has a screwdriver slot which in this case was slightly damaged by a screwdriver blade of the wrong size. The eccentric bushing is very hard to turn in some of the movements, so it is necessary to grind a screwdriver as shown in Figure 23. The blade is almost as wide as the slot in the bushing, providing good leverage. Even more important, the sides of the screwdriver blade have been ground parallel instead of tapered. The close fit which results makes all the difference in your being able to turn the eccentric without slipping and damaging the slot or the front plate. If an eccentric bushing needs to be replaced, it is available as part HER-2006.

ESCAPE WHEEL

Always check the escape wheel for bent teeth. Figure 24 shows the idea. A bent tooth can sometimes be straightened with a pair of smooth pliers.

Escape wheels also may have one or more teeth that have become shortened. You can usually hear a short tooth mislocking as the clock runs. When this happens, the short tooth lands on the impulse face instead of the locking face of the pallet unit. So much energy is lost recovering from this "lost" beat that the clock may frequently stop as a result.

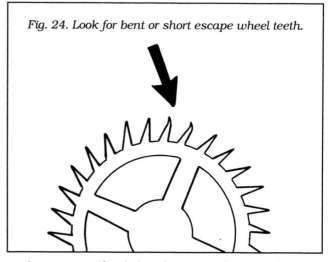

Fig. 24. Look for bent or short escape wheel teeth.

An escape wheel that has any short teeth must be replaced. Equalizing the length of all the teeth is not effective, since it changes the escapement's characteristics. Replace the wheel instead. Most tubular movements use part HER-22083.

OTHER THINGS TO CHECK

More defects will be highlighted throughout this book, but it would be well to call attention to the subject now. Look for any bent, rough, or damaged levers as soon as you can. When the movement is apart, you have your best opportunity to look at all the wheels and levers. The levers, especially, may be rough and a source of future problems. As shown in Figure 25, the hour rack and the quarter rack

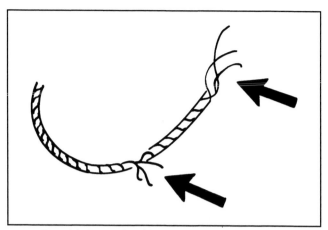

Fig. 26. Frayed cable ends or loose strands in the middle can cause a weight to drop.

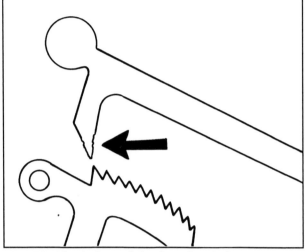

Fig. 25. Check the rack hook and rack for roughness that can cause chime and strike problems.

have critical surfaces at the place where the rack hooks rest against them. Roughness can cause a movement to chime on and on, without locking. Some repairers have attached lead weight to the rack hook, in an effort to make it lock in a positive manner. All that was needed was to carefully polish and clean the mating surfaces of the rack hook and rack. Doing a little polishing, where necessary, can save you a time-consuming job of troubleshooting.

Figure 26 reminds you to look for worn or frayed cables. You can order .060" diameter brass cable from Herschede. The correct length is 7'6". Remember that it is important to have to the right length, so the cable runs smoothly over the entire grooved portion of the winding drum. If you cut it too short, the clock will not run eight days. Allowing the cable to remain too long is even worse, because it will loop over itself at the back of the drum. The cable can foul the click, causing the weight to fall. Even one extra wrap of cable can be a problem. The last turn of cable can get jammed tightly against the edge of the clickwheel, reducing power enough to stop the gear train. Checking the cables early in the overhaul procedure will save you the trouble of trying to get new cable at the last minute.

It seems to make sense to change all the cables in a movement that is in your shop for a complete restoration, assuming the cables are at least four years old. Don't take a chance with old cables that could break in a year or two, dropping a weight.

Repair Tips from Howard Klein, Former Owner of Herschede

1. Check the leather pads on the hammers. Replacements are available from Herschede.

2. Keep the pulleys identified when repairing a movement. Because of the greater weight, the chime pulley may wear. If this worn pulley is installed on the time train, it may cause the clock to stop. Replace worn pulleys.

3. If either fly (fan shaft assembly) is too tight on the arbor, it will not spin after the gear train stops. This places extra stress on the gathering arbor front pivot hole, which may wear as a result.

4. Always check the pivot shoulders on the main arbor for roughness and wear.

4

REASSEMBLY & ADJUSTMENT

This chapter covers the most important parts of the Herschede overhaul, the reassembly and adjustment procedures. The part numbers and illustrations which follow are for the latest 9-tube movement. Remember, however, that the procedures are valid for three-weight, 9- and 5-tube Herschede models of almost any vintage. You will find it helpful to refer to the drawings and parts lists in Chapter 14.

Reassembly and adjustment are done together. After you install some of the parts, they must be adjusted for correct operation before the next parts are added. You may have to make adjustments to the gathering pallets, the cylinder, the self-adjusting lever assembly, the escapement, or other parts.

Following specific steps in putting the movement together is not enough; you need to understand the function of each part or assembly. Chime and strike mechanisms are especially difficult to adjust unless you know what each adjustment is supposed to accomplish. In this chapter, we will examine parts of the movement as we put them together.

Before you get started, there are a few things to do. You have to reassemble the main shaft assemblies, which you took down for cleaning. Do not attach the cables at this stage; put them on at the end of the job, just before you install the movement on the seatboard. Now lay out all the movement parts, and you are ready to proceed.

REASSEMBLY

It is a good idea to sort the wheels and arbors into three groups for chime, strike, and time before you begin. The main shaft assemblies are marked "C", "S", and "T". The chime second wheel is the large wheel with the beveled teeth. The strike second wheel is the pin wheel. And the time second wheel

is the center wheel, mounted on the long center arbor. To tell the strike and chime gathering wheels apart, look for the one with the pinion on the same end of the arbor with the wheel; it is the strike gathering wheel. The two warning wheels are the same, and have the same part number, 2626. If they aren't interchangeable on your movement, it is most likely because a modification has been made.

Figure 27 shows the first phase of the assembly procedure. The front movement plate is face down on the bench, and all the arbors have been loaded into it. Add the back plate, and carefully fit the pivots into their holes. As an alternative, you can assemble the movement by placing the arbors into the back plate instead, then installing the front plate. The internal movement view on page 75 looks at the movement from this perspective.

Do not install the fan shaft assemblies and the anchor at this time; they are added after the plates are together. Besides the gears, you must put in the maintaining shaft assembly. It fits in near the

Fig. 27. The arbors are shown loaded into the front plate, before the back plate is installed.

2106 strike gathering pallet
2109 chime gathering pallet
2655 hour bridge
2656 minute wheel (minute tube assembly)
2658 intermediate wheel
2660 hour rack
2661 quarter rack
2662 hour rack hook
2663 quarter rack hook
2665 lifting lever
2666 long lever
2667 self-adjusting lever assembly
2668 double lever assembly

Fig. 28. Major front movement parts, 9-tube model.

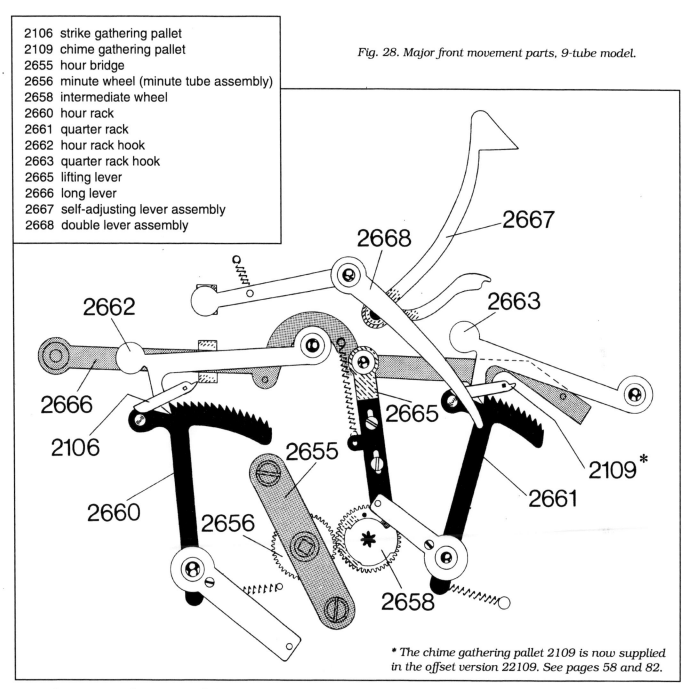

* The chime gathering pallet 2109 is now supplied in the offset version 22109. See pages 58 and 82.

spreader post, in the center of the movement, and acts upon the maintaining wheel, part of the time train main shaft assembly. Another part to add now is the hour strike shaft. Also look for the long lever shaft, a plain steel arbor turned smaller at one end. You can fit it into the front plate now, or insert it through the rear plate after the movement is assembled. Later in the procedure, you will mount the long lever on this arbor.

As you proceed, refer to Figure 28 for the location of the major front movement parts. Go to Chapter 14 to find any parts referenced in the text but not shown in Figures 28, 29, and 30. Without worrying about getting the wheels and pins synchro-

nized into final adjustment, fit the pivots into their holes. Leave the rear pillar screws finger tight. Add the hammer arm to the end of the hour strike shaft. Install the long lever (2666) onto the end of the plain arbor which you put in earlier. Make sure the lifting lever (2665) is placed to the right of the steel pin in the front plate. Add the hour rack hook (2662) and the quarter rack hook (2663). You can tell them apart easily because the quarter rack hook has a small pin on the end. Install the hour rack (2660) and quarter rack (2661). Next, put the two gathering pallets in place, but do not fasten them. In the latest 9-tube movement, the strike gathering pallet is part number 2106, and the chime is number

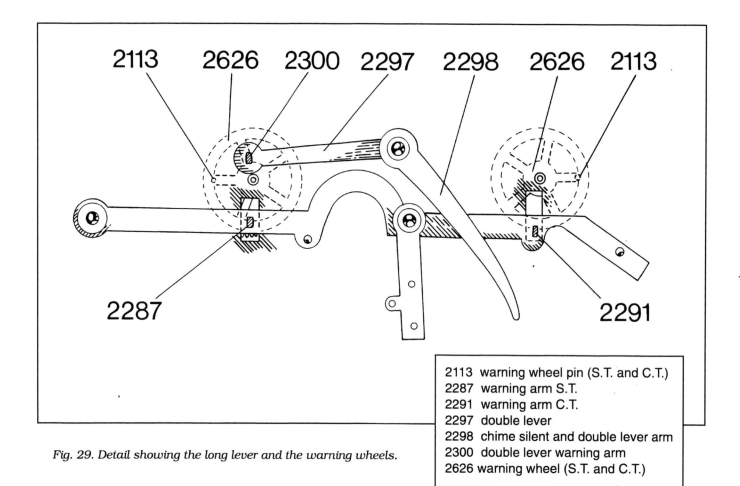

Fig. 29. Detail showing the long lever and the warning wheels.

2113 warning wheel pin (S.T. and C.T.)
2287 warning arm S.T.
2291 warning arm C.T.
2297 double lever
2298 chime silent and double lever arm
2300 double lever warning arm
2626 warning wheel (S.T. and C.T.)

2109, an offset pallet of different design. Your gathering pallets may appear to be identical, but you will know which is which after trying them on the arbors.

You are now ready to check the strike train. Figure 29 shows what to look for. The strike warning wheel is the dotted wheel on the left in the drawing. With your finger, turn the gears until the gathering pallet contacts the hour rack pallet pin, stopping the gear train. Look at the warning wheel pin. If it is not in an 8 or 9 o'clock orientation as shown in the figure, remove the gathering pallet and try another of its four possible positions on the squared end of the arbor. If none of the four positions gives you the correct warning wheel setup, you have to separate the clock plates enough to ease the warning arbor rear pivot out of its hole. With the gathering wheel disengaged, move the strike warning wheel to the correct position. Next, check the hour strike shaft assembly to make sure it is not in contact with the pin wheel when the strike train is locked. If the hour strike shaft is in contact with one of the pins, separate the plates enough to disengage the pin wheel. Keep the gathering wheel locked on the pallet pin as you move the pin wheel to clear the hammer tail. Using care, you can separate the plates

without bending any pivots or causing the wheels to fall out. Just remove the pillar screw on the affected corner, and loosen the other top screw and the lower screw on the strike side. Never force the parts into place. Check the strike train operation again, to be certain you have the 8 or 9 o'clock warning pin position, and that the hour strike shaft assembly is clear of the pins on the pin wheel.

Now move on to the chime train. The chime warning wheel is on the right in Figure 29. Run the gears through to the locking point, and with the gathering pallet resting on the quarter rack hook pallet pin, observe the warning pin. It should be at a 3 o'clock location as indicated in Figure 29. Try other gathering pallet positions, and if necessary separate the plates one last time to get the wheels in the right relationship.

You can proceed to install the rest of the front movement parts identified in Figure 28. Put on the self-adjusting lever assembly (2667), including the washer and wire fastener. Then add the double lever (2668) and its spring. The minute tube assembly (2656) can be pushed on, and then the hour bridge assembly (2655) with its two brass spacers and long screws can be installed. Continue by adding the intermediate wheel (2658). It must be put in the right position now, to ensure the clock will strike on the exact quarters. Mesh the minute wheel and the intermediate wheel so the lifting lever "drop off" comes at the quarter hours. Follow with the hour tube assembly (2657) and the intermediate bridge (2641), both of which are shown on page 74. Put the second shaft bridge assemblies, shown pages 6 and 7, on the front and rear. You will adjust them in a later step.

Figure 30 shows the crown wheel bridge (2154) with the crown wheel (2152) and bevel gear, to be installed next. You will adjust the assembly later. Put the hammer bar assembly, which includes the chime lever assemblies (2859) together with the cylinder assembly (2864) then proceed to mount them on top of the movement. Observe that the cylinder gear (2168) meshes with the crown wheel (2152).

Install the hour silent lever and chime silent lever, shown in Figure 8. Then refer to Figure 10 and add the anchor assembly, which includes the fork and the anchor bridge. Install the fan shaft assemblies and the fan bridges. Go back over the entire movement to be sure you have fastened all the levers in place with "s" shaped wires and the thin brass washers.

STRIKE AND CHIME OPERATION

At this point in the overhaul, it is worthwhile to study how the strike and chime trains work. If you have good knowledge of them, the rest of your adjustments will be easier to accomplish. As you read this section, refer to the drawings in this chapter and Chapter 14 as necessary.

Begin with the chime function. Lifting occurs with the pins on the intermediate wheel raising the lifting lever every quarter hour. A pin on the long lever pushes up the quarter rack hook to release the quarter rack. The quarter snail (mounted on the intermediate wheel) has four positions, one for each quarter. The distance the rack falls will determine how many notes are chimed. As the rack falls, the gathering pallet is released, and the warning pin moves to the warning arm (2291), shown in Figure 29. At the exact quarter hour, the warning arm drops as the intermediate wheel pin releases the lifting lever. The gathering pallet counts off rack teeth as the chime train runs. Locking occurs as the tail of the gathering pallet comes to rest on the quarter rack pallet pin.

Now for the hour strike function. Before the hour, a pin on the long lever raises the hour rack hook,

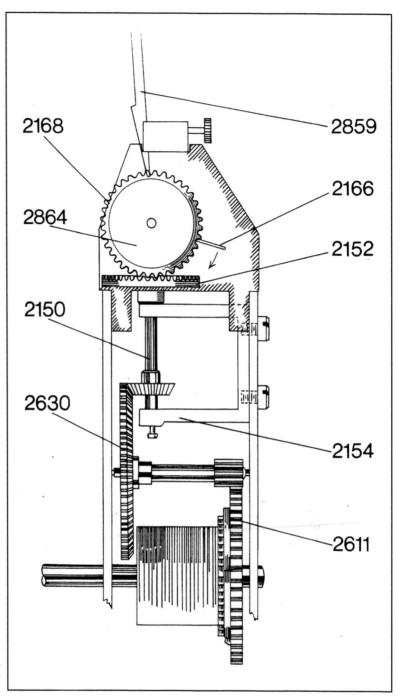

Figure 30. View from the chime side of the movement.

2150	crown shaft
2152	crown wheel 9T
2154	crown wheel bridge
2166	cylinder self-adjusting pin
2168	cylinder gear 9T
2611	C.T. main shaft assembly
2630	C.T. 2nd shaft assembly
2859	chime lever assembly
2864	cylinder assembly 9T

releasing the rack. (On the other three quarters, the lift is not high enough for release.) The rack tail falls to the hour snail, establishing the number of hours to be struck. As the rack falls, the gathering pallet is unlocked.

Strike warning occurs as the strike warning pin moves up to the double lever warning arm (2300), shown in Figure 29. Then, as the hour chime completes, the arm is lowered, allowing the pin to escape and the strike train to run. There are actually two warning arms for the strike train. During normal operation, the double lever warning arm (2300) functions. The warning arm for the strike train (2287) takes over for it only when the chimes have been silenced. When the chime silent lever is put on "silent", the double lever is pushed out of the way. It is then that the warning arm number 2287, attached to the long lever, takes over.

The double lever (2297) is operated by a coil spring. The spring keeps the double lever warning arm in the raised position whenever the movement is chiming or at chime warning. During this interval the clock cannot strike, because the double lever warning arm stops the warning pin. As the hour chime draws to a close, the quarter rack pallet pin contacts the chime silent and double lever arm (2298) and pulls against the spring pressure, moving the double lever warning arm out of the way. That is how the strike train is released to run. Be sure the spring is in good condition and properly attached. If it does not work, the double lever warning arm will not catch the warning pin. The movement will go to warning against the other warning arm (2287), but at the hour the clock will start chiming and striking at the same time.

Gathering Pallets

Figure 31 shows a group of Herschede gathering pallets, of slightly varying design. (The movement has only two gathering pallets, one each for chime and strike). A gathering pallet can be the cause of chime problems. A strain is placed on the pallet each time it stops the gear train. The square center hole can wear larger, especially if the fit is loose initially. Even if the part does not fail, it can wear so loose that the warning pin position becomes incorrect.

If the chime warning wheel pin position is correct, it will be at the 3 o'clock position as viewed from the front. The pin can then move more than a quarter revolution of the wheel before hitting the warning arm for the chime. If the locking action is later because of the gathering pallet, the warning pin may come to rest at a 6 o'clock position, very close to the warning arm. Sooner or later, the warning arm will jam against the pin and stop the clock.

Fig. 31. Herschede gathering pallets of slightly varying appearance, made in different years (see text).

If the owner forces the minute hand further, the warning pin may shear off. Even if the clock doesn't jam, the chimes may begin to stall because there will be almost no warning run.

A similar set of problems will occur if the strike warning pin does not come to rest at the 9 o'clock position. If the pin ends up at 6 o'clock, it interferes with the warning arm for the strike. A position near 11 or 12 will get you into trouble with the double lever warning arm. So watch out for a loose gathering pallet, or one which has been installed incorrectly. Even if the rest of the movement has been assembled the right way, a gathering pallet problem can stop the clock.

You should always carry several new gathering pallets in your tool kit, to use when the need arises. The part of the pallet which gathers up the rack teeth is critical. You may have to trim it slightly to prevent the pallet from pushing the rack ahead two teeth at a time. Equally important is the tail of the pallet. If it is too long, it will catch the pallet pin on the rack too early, before all the teeth are counted off. If the tail is too short, it will not lock on the pallet pin at all.

Self Adjusting Lever Assembly

The self adjusting mechanism, part number 2667, is shown in Figure 28. It automatically corrects or synchronizes the chime melody if the hands are moved or the clock is allowed to run down. Figure 32 highlights this area of the movement. The self adjusting arm catches the quarter rack pin at each hour. It holds up the quarter rack hook to prevent the gathering pallet and the rack hook from counting through and locking. If the chimes do not need correction, the self adjusting arm holds the rack hook up for only an instant. However, if synchroni-

Fig. 32. Self-adjusting lever and cylinder, 9-tube clock. This is the upper right portion of the front plate.

zation is needed, the arm holds the rack hook for a longer interval. The chiming simply continues until the cylinder self adjusting pin (2166), shown in Figure 30, comes around. This is an extra-long pin in the cylinder, long enough to lift the self adjusting lever and release the rack hook. The cylinder self adjusting pin is placed so that release occurs at the beginning of the correct hour chime note sequence. Four sequences of notes are then counted out for the hour chime. Unlike many other movements, the Herschede tubular bell movement will always chime the right number of notes on the first, second, and third quarters. It does this even if the chime sequences have been upset and the clock is playing the wrong musical notes for each quarter. On the hour, the self adjusting lever assembly takes over. The clock may chime as many as four extra note sequences before the start of the correct hour tune. A "sequence" or measure of notes is four for Westminster, six for Canterbury, and eight for Whittington chimes. Chime note sequences are covered in detail in Chapter 11.

Problems can result if the notch in the self adjusting arm is damaged in some way. It may fail to hold the quarter rack hook pin. There will not be any automatic chime correction if this happens. At each quarter and on the hour, the clock will chime the correct number of notes. However, once the sequence is thrown off for any reason, it stays that way. The notch in the self adjusting arm must be reworked, or the arm replaced.

You should verify that the chime self adjusting system works for all three chimes—Westminster, Canterbury, and Whittington. Almost by accident I did happen to run across a movement where the mechanism worked for two of the chimes but not the third. In some movements there is a slight dif-

ference in how the three chimes operate with respect to the self adjusting lever. The cylinder moves side-to-side from the left (Whittington) to the center (Canterbury) to the right (Westminster) as the chimes are shifted. Naturally, the cylinder self adjusting pin moves as part of the cylinder. The pin will contact the self adjusting lever at one of three places, depending on which chime is playing. If the acting edge of the self adjusting lever is not exactly level, the pin can give more lift on some chimes than on others. Bend the lever slightly to adjust it.

Cylinder Adjustment
Figure 33 shows a side view of the 9-tube movement. Refer also to Figure 30, page 17. The cylinder has pins which raise the chime lever assemblies (2859). The chime main wheel (2611) acts on the pinion above, to drive the second wheel (2630). This second wheel has beveled teeth which mesh with another bevel gear to rotate the crown shaft (2150). At the top of this vertical arbor is the crown wheel (2152). It drives the cylinder gear (2168) and the cylinder along with it. To adjust the cylinder it is necessary to disengage this drive system and rotate the cylinder independently to a known point.

Before you attempt this adjustment, take a moment to review an important point. Be careful not

Fig. 33. The arrow points to the gap created when the crown wheel is raised on its arbor to disengage it from the chime second wheel. The cylinder can now be moved by itself to correct the chime melodies.

to confuse the cylinder adjustment with the operation of the self adjusting lever assembly just covered. The self adjusting lever cannot synchronize the chimes unless the cylinder is installed correctly in the first place. It will keep returning the cylinder to the same incorrect starting point each hour, no matter how the minute hand is turned. On the first quarter Whittington chime, for example, the clock is supposed to play eight notes in order from highest pitch to lowest. If the cylinder has been installed the wrong way, the clock will still play eight notes, but they won't be the right ones. And most important, the clock will automatically return to this incorrect tune even if you manipulate the hands until the clock chimes correctly on the quarter hours.

You can adjust the cylinder when the clock is running on the test stand. With the dial off, the movement will be set for Whittington chimes. Turn the minute hand ahead to any quarter and allow the clock to chime. Do not expect the right notes to play, because you have not made any adjustments yet. Loosen the set screw on the bevel gear, and raise the gear up on the vertical arbor until it no longer meshes with the second wheel (Figure 33). Now grasp the cylinder and turn it independently. It turns clockwise from the viewpoint of Figure 30. Watch the hammers rise and fall. Look for the eight chime hammers to operate in sequence from right to left, viewed from the front. When you observe this pattern, which is the eight descending notes of Whittington, stop. Lower the bevel gear back into place, and tighten the set screw. The self adjusting assembly will take care of the rest by synchronizing the chimes on the next hour. The whole point is that you adjusted the cylinder in such a way that, when the chime train was locked at the end of its run, the cylinder was also at the end of a sequence.

Continue by checking the adjustment of the bevel gear. If it is meshed too deeply with the second wheel, the two will bind. If it is too shallow, there will be excessive slack in the gearing. The easiest way to adjust the depth is to work on the two second shaft bridge assemblies. One operates on the front pivot of the second arbor, the other on the rear. By turning the steel adjusting screws, you push the second arbor and wheel forward or back. Turning out both screws increases the endshake on the arbor. You must have endshake, but no more than necessary. Working the screws in conjunction with each other allows you to control the endshake and establish the depth of the second wheel and the bevel gear.

Alternate Method For Cylinder Adjustment

If it appears that the bevel gearing does not need to be adjusted, you may be able to eliminate that step and simplify the procedure by adjusting the cylin-

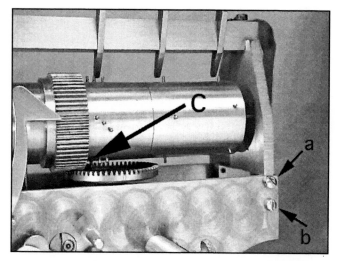

*Fig. 34. An alternate method of adjusting the cylinder is shown in these two photos. Above: the cylinder bridge screws at **a** and **b** are removed, as are the corresponding cylinder bridge screws in the back. One front and one rear cylinder bridge screw are removed from the other side of the cylinder assembly. The object is to disengage the cylinder gear from the crown wheel at **c**. Below: lift the end of the hammer bar and cylinder assembly. With the cylinder now disconnected from the rest of the chime train, it can be turned to adjust the chime melodies.*

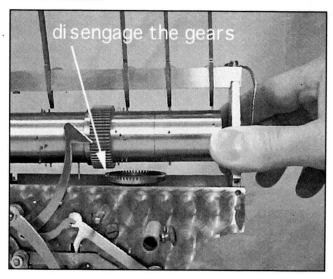

der alone. The goal is to move the cylinder independently to a new location where the chimes will play correctly. This can be done by disengaging the cylinder gear from the crown wheel. The cylinder can then be moved in one-tooth increments to a correct adjustment. From the procedure described in the previous section, recall that we want to set the cylinder to the end of a chime sequence when the chime train is locked at the end of its run. The chimes will then be correct and will be maintained that way by the self adjusting mechanism.

This alternate method of cylinder adjustment is illustrated in Figure 34. It may not save any time over the previously described method, but it has the advantage of leaving the bevel gearing untouched. It can be done with the weights on the clock and the tubular bell hammer cords hooked up. It can also be done on the bench with an assembled movement.

To begin, the clock is made to chime to the end of a sequence. Remove all the cylinder bridge screws except one front and one rear screw on the left (strike) side of the movement. Now raise the right side of the cylinder enough to disengage the cylinder gear from the crown wheel (at **c** in Figure 34). Turn the cylinder by itself until all the hammers are raised and then dropped in order from right to left. Now lower the cylinder until the gears mesh again. Reinstall the cylinder bridge screws.

Double check to be sure that the clock now chimes correctly, without any hammers left under tension at the end of a chime sequence. Similarly, it's important that the pins on the pin barrel are clear of any hammers when the chime warning occurs. If a finer adjustment must be made, you can always use the method described at the beginning of this section.

ESCAPEMENT ADJUSTMENT

The Herschede tubular bell movement has a deadbeat escapement whose pendulum swings once per second. The escapement is made up of the escape wheel, anchor, and supporting parts. Figure 35 shows a Herschede anchor and escape wheel.

Repairs to damaged escapements were covered in Chapter 3, "Basic Repairs". Sometimes, repair and adjustment are combined in a series of steps done at one time. That's the reason adjustment was included with repairs in Chapter 3.

There is very little adjustment that can be done. *The only adjustment on this escapement is the depth of the anchor, controlled by an eccentric nut on the front movement plate.* "Depth" means the distance between the escape arbor and the pallet (anchor) arbor. A shallower depth means the anchor and the escape wheel (Figure 35) have been brought closer

Fig. 35. Herschede anchor (top) and escape wheel.

together. The adjustment procedure has already been covered on page 12, where it was shown that the pallet arbor front pivot hole is located off center in an eccentric nut. When the nut is turned the hole is raised or lowered accordingly. Almost the only reason for changing the depth is to restore it to the original adjustment. You may find the slot in the eccentric nut has been marred by a screwdriver; someone tried changing the depth in an attempt to improve or experiment on the escapement.

Changing the depth in a deadbeat escapement affects primarily the "drop". Drop is the free motion of the escape wheel which occurs as one tooth is released by a pallet and another tooth is stopped by the other pallet. When the drops are equal on each pallet, the depth is correct; changing the depth increases the drop to one pallet as the other is decreased. *Do not touch this adjustment on a Herschede unless you are sure it is required.* If the escapement doesn't work properly, it may not be simply a matter of adjustment. There may be a problem with grooved pallets, bent escape wheel teeth, or dirt, that must be repaired first.

5

CHIME POINTS

The "chime point" is the position of the minute hand when chiming begins. The clock should start chiming exactly at each quarter, not half a minute early or a minute late. You may find that the clock chimes at the right time on some quarters, but not on others. Your customer has a right to expect these chime points to be reasonably accurate. What standard should you apply? The ideal is to have the clock start chiming 15 seconds before the hour to allow for the chimes and strike to span the hour nicely. Chiming should begin exactly at each of the three quarters. Practically speaking, a few seconds tolerance either way must be allowed on each quarter. The hour should start no earlier than 15 seconds early, and could be a second or two late.

Before you adjust the chime points, understand how chiming begins. Figure 36 shows the major parts. The process starts as the lifting lever extension drops off a pin. As explained in Chapter 4, the intermediate wheel raises and drops the lifting lever extension each 15 minutes. The lifting lever extension needs a fairly clean, sharp edge. A rounded corner will cause erratic results, with chiming for a particular quarter not always beginning at the same time. Older models have a different design for the lifting lever; there is no lifting lever extension, which means the lever is one piece and not adjustable. The bottom of the one-piece lever is curved instead of rectangular. The clock may stall if the intermediate wheel pins do not move smoothly as they push on the curved face of the lever. There can be a butting action as the lever is pressed against the steel pin in the front plate. It is most likely to happen at the hour, when the lifting action is higher than on the quarters. Polish out rough spots on the pins and the lever to relieve the problem.

Figure 37 shows the intermediate wheel (actually the full part name is intermediate shaft assembly). It meshes with the minute wheel on the front of the movement, and rotates counterclockwise as the clock runs. The assembly consists of the arbor, pinion, wheel with five pins, and the quarter snail. Why five lifting pins? Three of the pins are spaced the same distance out from the center of the wheel; these are for the three quarters. The other two pins are for the hour, which requires higher lift to release both strike and chime trains. The "extra" pin just helps create a smoother lifting action for the lifting lever extension. Older models do not have the extra pin.

By changing the mesh of the intermediate wheel and the minute wheel, you affect the dropoff point. There is only one setting which gives the best chime point. If you move the intermediate wheel only one tooth, the chime point changes by several minutes. Naturally, the intermediate wheel should be correctly set before you consider other adjustments.

ADJUSTING THE MINUTE HAND

The Herschede minute hand has a brass bushing which can be moved. Only a slight change will cause the clock to chime earlier or later, without affecting the clock mechanism in any way. The main problem seems to be finding a method for moving the bushing, which is very tight. You definitely need a special tool for this purpose—pliers won't do it. Figure 38 shows a tool made for me by Wilson Suggs, a Florida clockmaker. His tool is a tapered file tang in a plastic handle; the file tang fits through the minute hand bushing. The factory used a steel block with a square projection sized for the bushing. To use either type of tool, place the minute hand on the tool and then turn against the bushing.

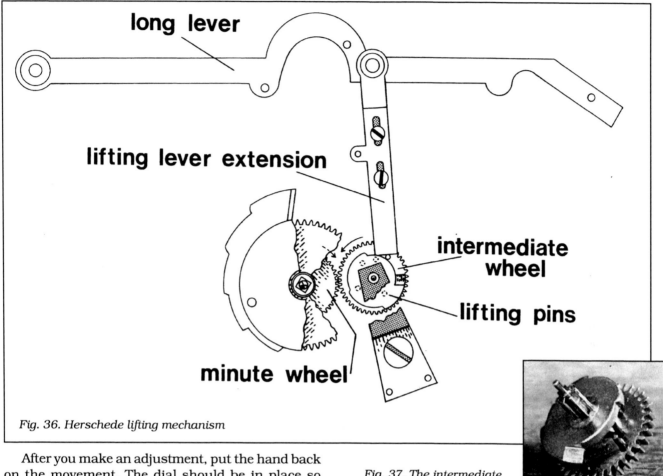

long lever

lifting lever extension

intermediate
wheel

lifting pins

minute wheel

Fig. 36. Herschede lifting mechanism

After you make an adjustment, put the hand back on the movement. The dial should be in place so you can accurately see the result of the adjustment. The best way to check the chime points is to remove the pendulum and allow the crutch to tick rapidly. This will simulate the actual running of the clock, and save you time waiting for the chiming points. Turning the minute hand to the chime point is less accurate, because you may turn it too far.

If all four chime points are where you want them to be, you are finished with the adjustment. Often, however, you will discover that the clock is further

Fig. 37. The intermediate wheel, or intermediate shaft assembly, part 2658

out of adjustment on the first quarter, for example, than on the third. Or one quarter may be a little early, and another late. Use your tool to find the best bushing position, which gives the most satisfactory all-around adjustment. If you cannot settle for the result, you must go to the second adjustment procedure, which involves the intermediate wheel.

INTERMEDIATE WHEEL ADJUSTMENT

Before you attempt to adjust the intermediate wheel, make sure you have obtained the best possible results by moving the minute hand bushing as described above. Decide which quarter or quarters are too far out of adjustment to live with, and concentrate on them. With the dial off, turn the minute hand to the "offending" quarter and identify the pin you want to adjust. Now remove the hour tube assembly and the intermediate bridge. But before you take out the intermediate wheel, use a felt tip pen to mark the teeth which mesh between this wheel and the minute wheel. This step saves you a trial-and-error job when you reinstall the intermediate

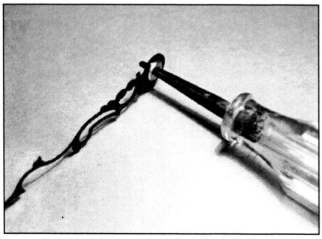

Fig. 38. Adjusting the minute hand bushing

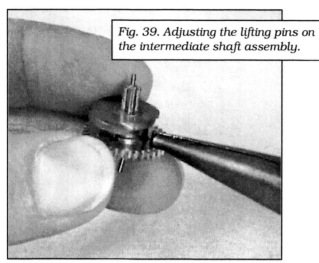

Fig. 39. Adjusting the lifting pins on the intermediate shaft assembly.

slightly in the right direction. You will see that if you would bend the pin toward the lifting lever extension, the dropoff will occur later; if you bend the opposite way, chiming begins earlier. After you make an adjustment, you must install the intermediate wheel and put the intermediate bridge back in place to accurately observe the new chiming point. The practical side of this adjustment is that you must not bend the pin too much or too often. It will come loose or fall out, and you will be in a bad position if you are attempting this adjustment during a service call.

Another way to make this adjustment is to carry a replacement intermediate wheel in your tool kit. Try the new wheel instead of taking a chance on bending the pins on the old one. If you use a new wheel, you must find the right meshing of the gears and then adjust the minute hand bushing for the best possible result. Whether you plan to use it or not, you should carry a spare intermediate shaft assembly in your tool kit.

wheel later. With the wheel still in place, move the minute hand back and forth across the chime point, to aid yourself in figuring out which way to bend the pin. Remove the intermediate wheel, grasp the pin with pliers as shown in Figure 39, and bend it

6

ADJUSTING THE HOUR SILENT LEVER

The hour silent lever is part number 2673 for both the 5- and 9-tube movements. It is located on the left front part of the movement. It stops the hour rack from moving whenever the hour strike is silenced. The selector pin pierces the dial at the slot near the numeral 9. Only two positions are possible: up for strike, or down for silent. It is a simple device, and we can easily take it for granted on the tubular bell movement. That is, until something goes wrong with it. This chapter covers the hour silent lever operation and adjustment.

You won't encounter a problem with the hour silent lever very often, but you should be ready when it does happen. The problem is particularly hard to correct because you cannot observe it directly. When you remove the dial to have a look, the problem seems to go away, only to return after the dial is put back on again.

The customer's complaint runs something like this. His 9-tube Herschede chimes the quarter hours correctly. But the hour strike counts properly only through 9 o'clock. At 10, 11, and 12 o'clock, it strikes

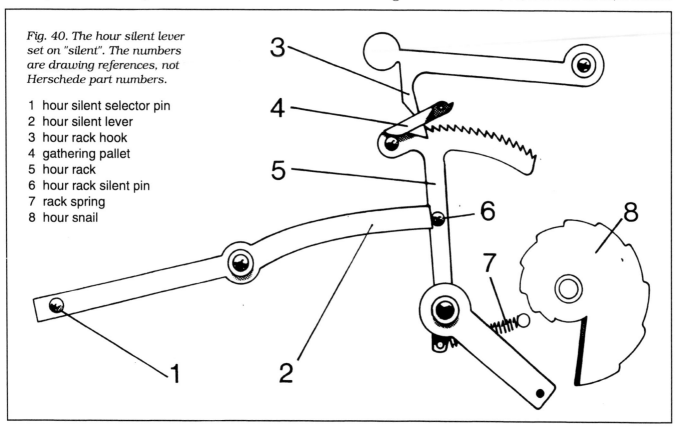

Fig. 40. The hour silent lever set on "silent". The numbers are drawing references, not Herschede part numbers.

1 hour silent selector pin
2 hour silent lever
3 hour rack hook
4 gathering pallet
5 hour rack
6 hour rack silent pin
7 rack spring
8 hour snail

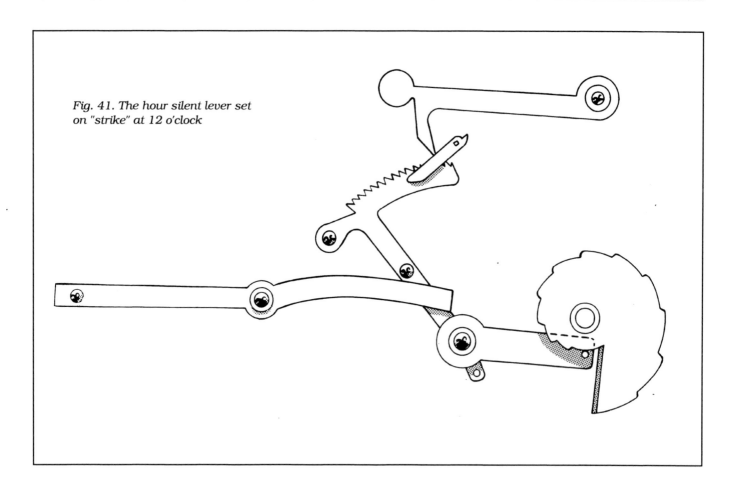

Fig. 41. The hour silent lever set on "strike" at 12 o'clock

9 each time. He is certain that he is not imagining it. The strike weight is lagging somewhat behind the other two weights. In some clocks, the strike operates correctly only through 7 or 8. I have seen a movement which struck a maximum of 5 notes on the hour.

First, you have to check a few points to convince yourself that you do have a problem with the hour silent lever. Refer to Figure 40 to identify the hour silent lever and the nearby parts. Remove the dial and check the action of the hour rack (5 in Figure 40). The rack spring (7) must be hooked onto the rack spring stud, and the spring must not be distorted or damaged. The hour rack must move freely to the 12 o'clock position without hesitation. Move the minute hand slowly to the hour, watching the strike warning action. The lifting lever assembly (shown on page 74) must raise the hour rack hook (3) high enough to give the rack clearance to move. Having eliminated these causes, you can be reasonably sure the hour silent lever itself is at fault.

Before you attempt to adjust the hour silent lever, take a moment to study how it works. The lever stops the hour rack from moving to any of the hour positions indicated on the hour snail (8). Figure 40 shows the lever in the "silent" position. The far right-hand side of the lever rests against the hour rack

silent pin (6) on the hour rack. As the hour silent lever is moved up to the "strike" position on the dial, the right-hand side of the lever moves downward. Figure 41 shows the clock in the "strike" mode, about to strike 12. The hour silent lever is clearly out of the way of the hour rack silent pin, and will not interfere with the rack movement at any hour.

Figure 42 illustrates the problem which concerns us. The lever is set on "strike", and the movement should be ready to strike 12. But the right-hand side of the lever is too high, and the hour rack silent pin hits it as the rack moves toward the 12 o'clock position. Although the hour silent lever limits the rack travel to 9 in our example, the critical hour may be as low as 5 or 6. The low-numbered hours are struck correctly, but the rack cannot go beyond the critical point.

With the dial back on, you cannot see the hour rack or the right-hand side of the hour silent lever. With the dial off, the entire problem seems to go away. This is because it is the slot in the dial which supports the hour silent lever and determines its position. The solution to the problem lies with this slot and the hour silent selector pin (1) which goes through it.

By comparing Figures 41 and 42, you can see that the corrective action you must take is to move

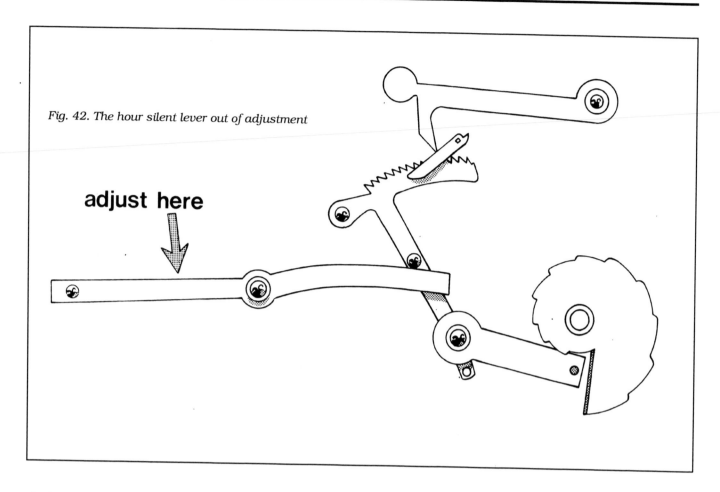

Fig. 42. The hour silent lever out of adjustment

adjust here

the lever so that its right-hand side is further down, as in Figure 41, where it will not interfere with the hour rack. You can accomplish this by bending the lever at the adjustment point indicated by the arrow in Figure 42. Grasp the lever with pliers and twist it slightly. This twisting action will cause the selector pin to point slightly downward. When you put the dial back on, the lever will move further toward "strike" than it did before. Test the strike and listen for a full count of 12. If it only counted up to 9 before the adjustment, and now counts to 11, for example, remove the dial and bend the lever slightly more.

The difficulty in making this adjustment is that once you put the dial back on again, you cannot see your adjustment in action. Three or four trials might be necessary. If the clock strikes 12 with the dial in place, you have been successful.

7

ADJUSTING THE CHIME SILENT LEVER

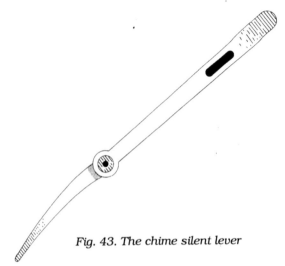

Fig. 43. The chime silent lever

If there is an adjustment problem in the chime silent lever of a 9-tube movement, it will be reflected in one of two customer complaints. One of these is that the silent lever does not do its job—the clock still chimes. The other complaint goes something like this: "The clock has always chimed perfectly. But then I put it on 'silent'. When I moved the indicator back to "chime", the clock did not start to chime, and hasn't chimed since." You can look for a chime silent lever problem during a routine check. All you have to do is silence the clock, verify that it does not chime, and then make sure it resumes chiming when you turn it back to "chime".

Let's take a look at the mechanism first, and then go on to cover the adjustments. The chime silent lever is pictured in Figure 43. The lever is carried on a stud on the front of the movement. On the upper right corner of the dial, there is an indicator marked "Chime" and "Silent" as shown in Figure 44. Behind the indicator is a cam with an offset pin (the silent shifting cam assembly). The pin operates in a slot (shown in black, Figure 43) in the chime silent lever. When you move the indicator hand, the pin pushes the lever.

Figure 45 shows the front movement parts which are important for our discussion, with the clock on "silent". The chime silent lever (1 in Figure 45) touches the quarter rack pallet pin (6), preventing the quarter rack (5) from moving. Since the quarter rack is held stationary, the gathering pallet (4) remains locked on the quarter rack pallet pin and there can be no movement in the chime train gears. At the hour, the quarter rack hook (3) is raised by the lifting lever assembly (not shown) and held there from then on by the self adjusting lever assembly (2). The silencing action is the simple result of the fact that the quarter rack does not move.

When the indicator is switched back to "chime", the chime silent lever pivots clockwise. The lever moves away from the quarter rack pallet pin, leaving the rack free to move. The clock will begin to chime as soon as the indicator is turned, unless the minute hand is at the warning position. At the next hour, the self adjusting lever assembly will make any necessary corrections to the chime note sequences automatically.

It is possible for the chime silent lever itself to be out of adjustment. The lever, as shown in Figure 43 is actually made up of two steel levers and a center bushing. The parts are firmly staked together to form a unit. Each lever assembly must be individually fitted for the movement in which it will operate. If, for one reason or another, the chime silent lever is not matched to the movement, its two halves may combine to form a unit which is slightly out of shape. The mismatching may be a result of previous repair attempts on the original lever. Perhaps someone replaced the lever instead, but did not adjust the new part. Two things can happen. The lever may not stop the rack, allowing the clock to chime when the indicator points to "silent". Or just the opposite may occur. Instead of just stopping the rack, the

Fig. 44. Chime silent indicator on the Herschede dial.

Fig. 45. Chime silent lever set on "silent" The numbers are drawing references, not Herschede part numbers.

1 chime silent lever
2 self adjusting lever assembly
3 quarter rack hook
4 gathering pallet
5 quarter rack
6 quarter rack pallet pin
7 double lever assembly
8 double lever warning arm

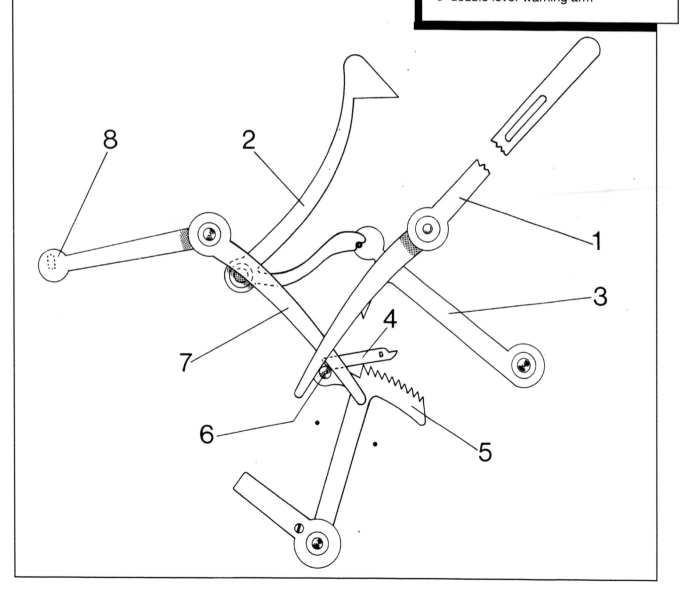

lever actually pushes the rack to the right. This can jam the chime train.

If the left half of the lever is bent too far down in relation to the other half, the unit will push the rack as pictured in Figure 46. This happens when the chimes are put on "silent". As the rack moves, the

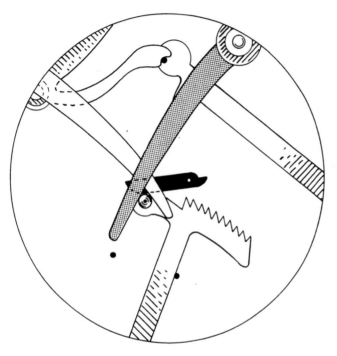

Fig. 46. Chime silent lever out of adjustment, pushing the quarter rack

quarter rack pallet pin is pushed along underneath the gathering pallet. Figure 47a shows what happens. Compare this locking action with the way the pallet and pin should lock as indicated in Figure 47b. This pinpoints the problem. When the indicator is switched back to "chime", the quarter rack still remains stuck because the gathering pallet will

not release it. In Figure 47a, the gathering pallet presses down on the pallet pin with all the force of the gear train. If the pallet pin had not been pushed so far under the gathering pallet, it would have slid off the curved tip of the gathering pallet (Figure 47b) and chiming would have resumed.

The gathering pallet and quarter rack pallet pin can be separated again just by touching them. But if a clock with a defective chime silent lever is silenced when the gear train is at warning, a more serious jam-up may occur. Figure 48 shows this "worst case" situation. First, the movement goes to the warning position. The gathering pallet is released, and the warning pin moves to the warning arm for the chime train. (The warning pin and warning arm are not shown.) As the indicator is moved toward "silent", the chime silent lever pushes the rack all the way over to the stop pin. The quarter rack pallet pin pushes the double lever assembly (Figure 45, part #7) counterclockwise as far as it can pivot. At the chiming point, the gathering pallet turns around and locks hard on the last rack tooth. The chime train is now jammed. The gathering pallet pushes on the rack, and the quarter rack pallet pin puts more pressure on the double lever, which is already "bottomed out". You may have to remove the chime silent lever and the double lever from the movement in order to release the jam.

Next, consider what happens if the left half of the lever is bent too far upward in relation to the other half. This is the opposite of the problem just described. Instead of the rack being stopped or moved to the right, it is allowed to move left, in its operating direction. The chime silent lever does not come into contact with the quarter rack pallet pin when the indicator is turned to "silent". The clock chimes. If, however, the chime silent lever is only very slightly out of adjustment in this direction,

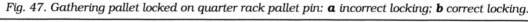
*Fig. 47. Gathering pallet locked on quarter rack pallet pin: **a** incorrect locking; **b** correct locking.*

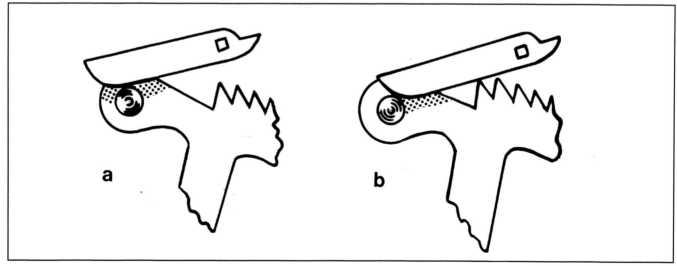

another problem can happen. The rack moves just the equivalent of a half rack tooth. The chimes are silenced, because the gathering pallet is still just barely held by the quarter rack pallet pin. But the pin does move just that half-tooth distance with the rack. The pin pushes the double lever assembly slightly. This causes the double lever warning arm to get in the way of the strike warning pin. At the hour, the strike train starts up (as it should if you haven't silenced it) and you get a clicking sound along with the strike. This is the sound of the strike warning pin glancing repeatedly off the double lever warning arm. It is also possible for the chime silent lever to be out of adjustment in such a way that the rack moves a little further than the half tooth. You still have chime silencing, but the strike doesn't work at all, because the double lever assembly moves enough to stop the strike warning pin.

Now you are ready to adjust the chime silent lever. To do this, you have to reposition the two parts of the lever so they form a unit of exactly the right shape. The unit must be able to stop the quarter rack, without pushing it. After you make the adjustment, stake the lever tightly together again. The adjustment will be small, and that is why patterns or measurements are of no use. Each chime silent lever is individually fitted to a movement. Unfortunately, this adjustment is difficult to accomplish in a customer's home, with the movement in the clock case. You cannot see around or over the dial to see whether the lever is just stopping the rack, or may be pushing it. When you remove the dial, you take away the indicator unit with the offset pin which positions the chime silent lever. The lever "flops" and you cannot tell how it will work when the dial is on again. It is best to adjust and check the lever when the movement is set up on a test stand. If this is not practical, I suggest the following procedure to be used in the customer's home.

First, determine which way you must bend the chime silent lever. If the clock will not silence, or has developed the related strike problems just described, remove the lever from the clock and reshape the unit so the left half will be lower. This will bring it closer to the quarter rack pallet pin. If the chime train has been jamming or failing to go back on "chime", bend the left half of the lever so it will be higher, and further away from the pallet pin. Remember, this is not actually bending, but sliding

the lever with respect to its center bushing. Before tightening the unit up again, you must be sure it works properly. Repeated trials may be necessary, with the dial, of course, having to be removed after each adjustment.

Finish up by thoroughly checking the operation of the chime silent lever. Make sure you get silencing, and that the clock resumes chiming when you turn the indicator to "chime". Listen for the hour

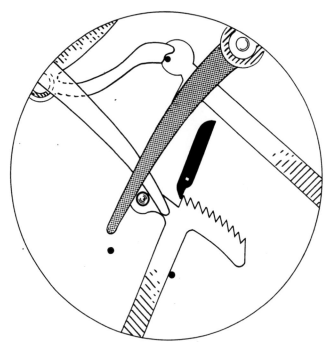

Fig. 48. Chime train jammed because of the chime silent lever

chime, which should still work if you haven't silenced it. One thing very hard to observe is whether the chime silent lever is pushing the rack to some extent. Look in the glass side panel of the clock. As you move the indicator to "silent", watch the chime warning wheel. If the chime silent lever is pushing the rack, you will see some movement in the wheel. This motion is caused by the quarter rack pallet pin sliding underneath the gathering pallet as in Figure 47a. The pallet is slightly lifted, imparting a backwards motion to the gears. Anything more than the barest "quiver" in the warning wheel means you are headed for trouble; the chime silent lever is pushing the rack. Remember, the danger is that the gathering pallet will hold the pallet pin instead of releasing it when the clock is put back on "chime". If you are in doubt, readjust the lever.

8

THE DIAL

The Herschede dial is the focal point of the tubular bell clock. Each dial is heavily constructed, and has many individually finished and fitted parts. From the repair standpoint, the dial can be the source of certain problems. To begin with, you have to know how to remove and install the dial correctly, because the clock will not work if you err. There is simply a knack to putting it on the right way. It takes practice, most of which is obtained under the customer's watchful eye!

Figure 49 shows the 9-tube clock dial. There are several features we should identify now. The chime indicator (b) is at the upper left part of the dial. The chimes are Westminster, Whittington, and Canterbury. Near the numeral "9" is the hour silent lever (a), which is used to silence the strike. The chime silent indicator (f) is at the upper right, and operates independently of the hour silent lever. There is a seconds bit (g) under the numeral "12" and above the center hole. At the top of the dial is the moon disc (c), part of a separate mechanism operated by the clock movement.

REMOVAL

To take off the dial, remove the minute, hour, and second hands (i, j, and h) first. Then reach behind the dial to release the spring loaded fasteners which hold the dial in place. There isn't much room to get your fingers behind the dial, and I've had sore fingers from trying to pull up on the brass "buttons" to release them. When they are all free, carefully pull off the dial. Don't drop it—it's heavy.

An overall dial check is in order. Look at the chime indicator. It should move freely but not be loose. The indicator should always be turned clockwise to counteract the tendency to unscrew the assembly; be sure to tell the customer. If you need to tighten

the indicator nut, cushion the pliers so you will not damage the finish. Taking a precaution like this is not being finicky. It's showing regard for an expensive dial. Check the chime silent indicator, and tighten it if it is loose. It operates from "silent" to "chime" and back again, not in a complete circle.

MOON DIAL

You have two dials to worry about. One is the clock dial, shown in Figure 49. The other is the moon dial (c and d in Figure 49 and Figure 50). Figure 51 shows a rear view of the clock dial, featuring the moon dial mechanism. Remove the dial and look at the back of it, to see how it works. Move the moon shift lever assembly to the left. The shift lever and pin slide over the teeth on the disc, but the disc does not move. This action "cocks" the mechanism for the next change (changes come every 12 hours). Now move the assembly over to the right. The shift lever pin engages a tooth on the disc, and the disc clicks as it jumps one space counterclockwise as you are viewing it in the illustration. The moon dial spring assembly holds the disc in place between changes. The clock operates the moon dial assembly by means of the moon eccentric disc, a brass disc on the hour tube. On this disc is an offset pin which revolves once in 12 hours with the hour tube. As it moves, its eccentric movement gives the back-and-forth motion to the moon dial mechanism as it pushes on one side of the moon lever assembly slot and then the other.

MOON DIAL ADJUSTMENT

If the moon dial does not work when installed on the clock movement, remove the clock dial and look at it again carefully. The very first thing to establish, before you do anything, is that the moon spring

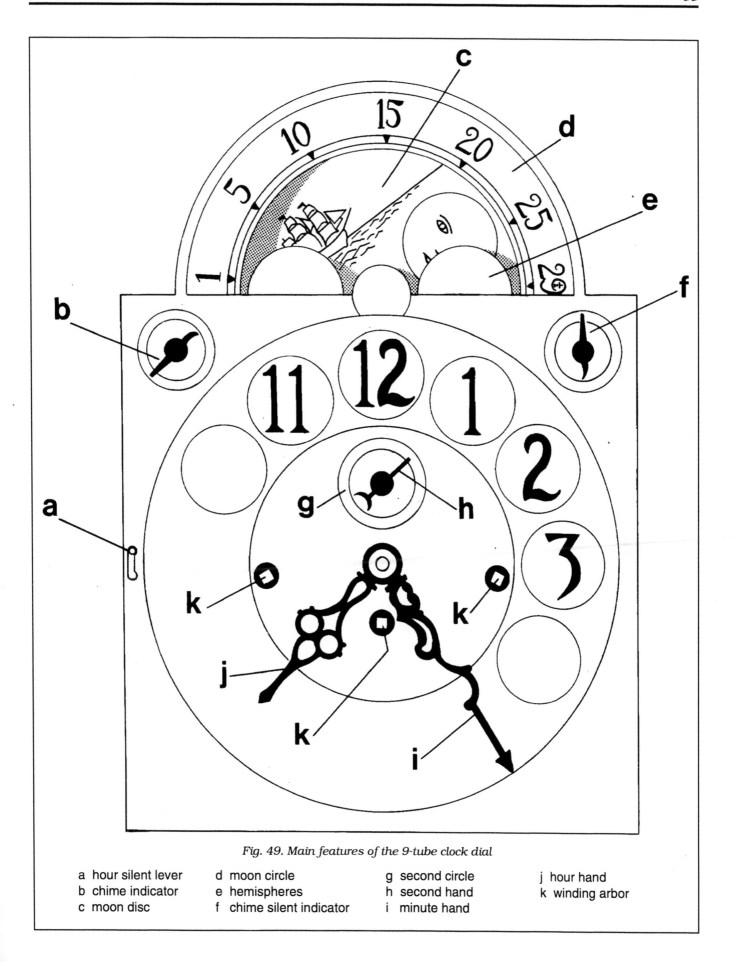

Fig. 49. Main features of the 9-tube clock dial

a hour silent lever	d moon circle	g second circle	j hour hand
b chime indicator	e hemispheres	h second hand	k winding arbor
c moon disc	f chime silent indicator	i minute hand	

assembly is not stuck behind the moon disc. If it is stuck, moving the disc will cause the spring to scratch the finish off the moon picture, ruining it. Carefully move the spring assembly back onto the teeth of the moon disc. Now move the moon shift lever assembly from side to side slowly. If you have to move it noticeably further to one side than the other to get it to operate, then the moon spring needs to be adjusted. The idea is to equalize the travel on each side. The screws holding the spring to the clock dial can be loosened, and the spring moved up or down. This adjustment permits you to control the rest position of the moon disc. Try a few spring locations and you will be able to equalize the motion of the shift lever assembly to each side. When you put the clock dial back on, check the moon dial assembly to be sure it works.

PUTTING THE CLOCK DIAL BACK ON

There certainly is a knack to putting on the Herschede tubular bell clock dial. All it takes is a method. Here is mine.

Before you even pick up the dial, locate the snail on the front of the movement. You can move the snail without affecting the gearing. Move it now until the offset pin on the brass disc is pointing downward. The reason for doing this is to place the pin out of the way of the moon lever assembly. You will have an easier time of installing the dial if these parts do not interfere with each other.

Next, pick up the dial. Move the chime indicator to "Whittington". This will make it easier to get the cylinder shift lever assembly (part of the movement) onto the 3-chime shifting cam assembly (part of the dial). The cylinder has three positions, and is spring loaded to the left. This is the position we are going for, because it corresponds to the flattest, "no lift" position of the cam. It's simply easier to get the parts together this way.

You are almost ready to put on the dial. Center the moon lever assembly so it is not pushed to one side. This is very important. The dial will still go on with the moon lever assembly jammed on the brass disc on the hour tube, and the clock will stop within a few hours if you don't correct the problem.

Fit the dial onto the clock by easing it on in stages. In general, it seems to go on best in the following sequence:

1. Place the hour silent lever in position in the "strike" slot in the dial, which is near the numeral "9".

2. Put the cylinder shift lever assembly on top of the 3-chime shifting cam assembly.

3. Continue to ease the dial further on.

4. Fit the offset pin for the chime silent lever into the slot in the lever.

Fig. 50. The moon dial.

5. Seat the fasteners which lock the dial onto the movement pillars.

6. Look in the side glass panels of the clock, and pull gently on each of the four corners of the dial, to make sure that the dial is all the way on.

SETTING THE MOON DIAL

With the dial in place, you are now ready to install the hands. After you verify that the strike train is locked (not ready to strike), put on the hour hand and turn it slowly clockwise. That's right, turn the hour hand. It will move only the snail and the disc with the offset pin. If the moon dial assembly is in good order and the clock dial is on properly, the moon disc will click ahead one space per revolution of the hour hand. Two or three turns are enough to confirm the moon dial operation. Naturally, if the moon dial does not work, remove the dial and reinstall it carefully. Recheck the moon dial mechanism, and look for jamming of the moon lever assembly.

Install the minute and second hands. Assuming the weights are in place and the clock is otherwise set up and operational, go through some checks. Try silencing the strike, then the chime, and both together, moving the minute hand ahead after each change. Turn the chime and strike back on and listen to the clock as you select each of the three chimes in turn. The customer usually likes to watch this whole procedure, ready to let you know if the clock strikes the wrong count! You really should check all these points to do a thorough job, and if you run into a problem it may be related to the clock dial in some way.

To set the moon dial, first determine the date of the most recent full or new moon. Then turn the moon dial ahead to the full or new position, whichever it is you have selected. Finally, count how many days have elapsed since your known point (full or new). Just reach over the top of the dial and click the moon disc ahead clockwise, twice for each day

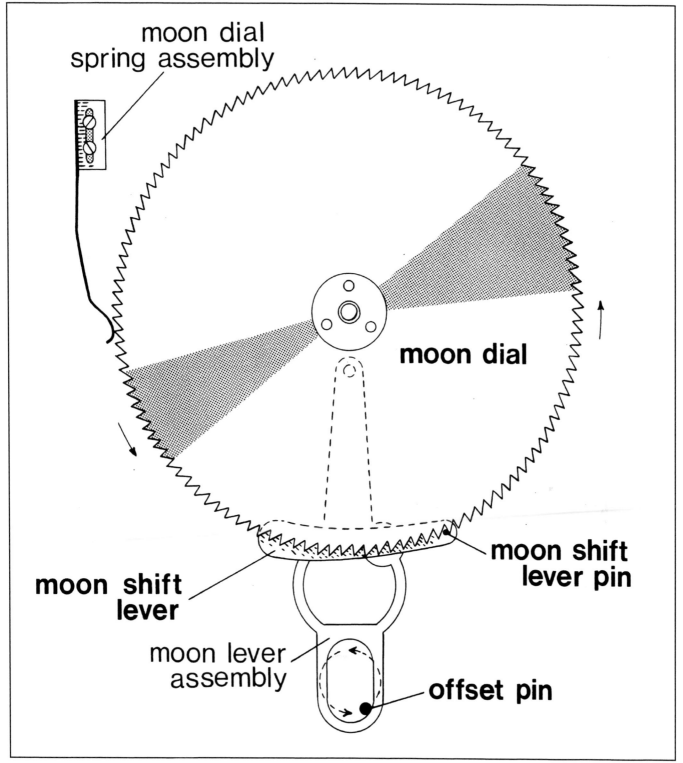

Fig. 51. Rear of clock dial, showing moon dial mechanism

you are adding. Explain to your customer how to set the moon dial. Most clock owners don't really want to become too involved in learning about the phases of the moon. They are often rather overwhelmed by the clock, and perhaps a bit afraid of it. Keep your instructions nontechnical and brief.

9

THE 5-TUBE MOVEMENT— THREE WEIGHT

The 5-tube Herschede movement has not been as popular overall as the 9-tube version. As the 1970's surge in the demand for grandfather clocks brought with it a renewed interest in triple chimes and extra features, the manufacturers all took part in a trend toward larger, fancier clocks. Herschede clocks remained the most elegant in the industry. Perhaps it is only natural that customers, once they had decided on buying a large Herschede, wanted the "better" version of the clock— nine tubular bells instead of five. Until late in the decade, the other American clock companies were limited to the 5-tube movements available from Germany, in addition to rod chime models. Herschede offered the only 9-tube grandfather clocks the American public saw on display.

If you learn to repair the 9-tube Herschede clocks, you will be able to take care of the 5-tube clocks as well. The 5-tube version plays only Westminster chimes instead of three melodies. There is no chime selector, and of course the number of bells is only five; four are for the chime, and one for the hour. The clock plates are rectangular instead of flared at the upper part. The flare is not necessary because the cylinder is shorter for five bells. Although the difference in shape of the 5- and 9-tube movement plates makes them look different from each other, they really are very much alike. Most of the wheels and levers are identical in design to the 9-tube parts. The repair and adjustment procedures are basically the same.

In this chapter, we will look at three versions of the five tube, three-weight movement. One is a newer movement. Another is an older version with

a different hammer arrangement. The third is a completely different Herschede tubular bell movement from the 1920's.

5-TUBE MOVEMENT—(NEWER)

The most recent version of the movement is shown in Figure 52. This rear view emphasizes the upper part of the movement and the chime and strike mechanisms. Most of the differences compared to the older model lie in this area. The front of the movement is much the same, new or old.

Notice that the five hammers are arranged in a row across the top of the movement. The four on the left, in this rear view, are chime levers (2859). The hour strike lever (2221) is on the far right. Linkage runs from the hour strike shaft assembly (2645), shown in cutaway, up to the hour strike lever. This means that the five tubular bells are hung in a row from the tube rack, in the back of the clock cabinet.

The cylinder gear (2167) is mounted on the cylinder shaft (2157), rather than being wrapped around the cylinder (2160). The banking set screw (2023) can be adjusted to permit more or less lift from the pins on the pin wheel, but is normally left alone.

A major difference with the 9-tube clock (and the older 5-tube) is the placement of the hour strike tube alongside the chime tubes. This means that instead of a hammer arm which moves across the back of the movement, we have the linkage described above. The modern linkage setup is not necessarily an improvement over the other type; it is merely a

(text continued on page 39)

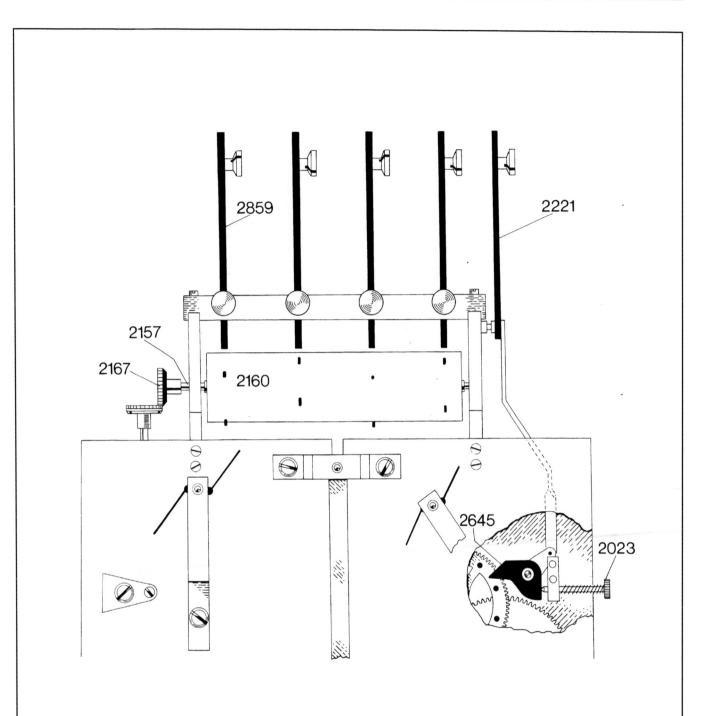

Fig. 52. Rear view of a newer Herschede 5-tube movement, showing the hammer arrangement

2023 banking set screw
2157 cylinder shaft
2160 cylinder 5T
2167 cylinder gear 5T
2221 hour strike lever
2645 hour strike shaft assembly
2859 chime lever (four)

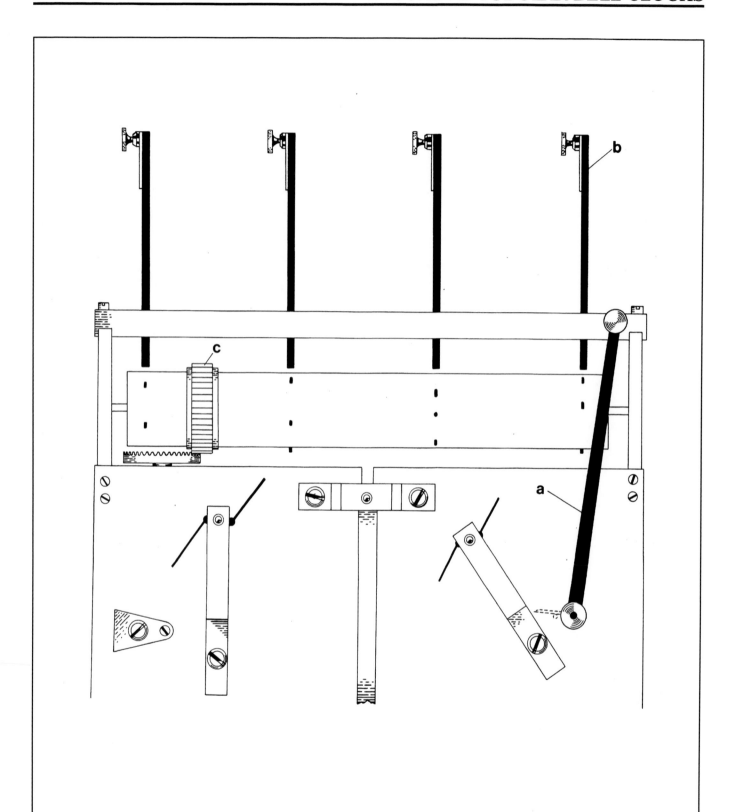

Fig. 53. Rear view of an older 5-tube Herschede movement

  a hammer arm (for hour strike)
  b chime lever (four across top of movement)
  c cylinder gear

...text continued from page 36

means of getting the five tubes in a row for the sake of appearance.

5-TUBE MOVEMENT—OLDER VERSION

The older 5-tube movement in Figure 53 is also shown in a rear view. It was drawn from a Herschede with a 1919 patent date stamped on the quarter rack, on the front of the movement. Ironically, many of the details remind us more of a 9-tube movement than a newer 5-tube.

Just like the 9-tube, it has a hammer arm across the back of the movement. The hour strike tube is hung off to the side instead of in a row with the chime tubes. The hammer arrangement is basically the same as the 9-tube, except for the fact that there are four chime hammer levers across the top of the movement instead of eight. The cylinder gear is mounted on the cylinder, not on the cylinder shaft.

Design of certain parts does vary from the newer 5- and 9-tube clocks. The hour strike shaft, or hammer tail, is a straight piece instead of the adjustable assembly in the newer clocks. The cord adjusters are clamping devices instead of the newer "wraparound" type. The rack springs and double lever spring are flat instead of coil springs. The moon eccentric is a cam instead of a flat disc with an offset pin. On the intermediate wheel, there are only four pins instead of five. And the lifting lever is a single piece, without an adjustable extension, designed with a curved tip instead of a rectangular one.

Many of these differences are minor, but you should be aware of them, to the extent that they affect repair and adjustment procedures.

FIVE-TUBE MODEL (1920's)

This section covers an example of a Herschede tubular bell movement quite different from the other two movements in this chapter. The plate size is 152mm wide by 136mm high. The back plate bears the Herschede crown symbol and name. Eight patent dates from 1914 to 1921 are stamped above the symbol. The plate is stamped with the serial number 82250. A list of Herschede serial numbers places this in 1928. The distinctive Herschede plate finish is present, and some of the parts look like those from more common Herschede models.

The movement does not beat seconds. The pendulum is four inches shorter than other Herschede pendulums. One wheel has been changed in the time train. The third wheel, just above the center wheel, has been placed closer to the center arbor. The dial is a less expensive one with painted instead of raised numerals and a blanked-off seconds bit covered with scroll work. Taking these things into account, it seems that Herschede manufactured this movement for a less expensive, smaller floor clock.

The chime train is the locking plate type instead of rack and snail. The strike is rack and snail; locking occurs on the round, notched gathering pallet. The usual dual rack and snail system is not used in this movement.

...continued next page

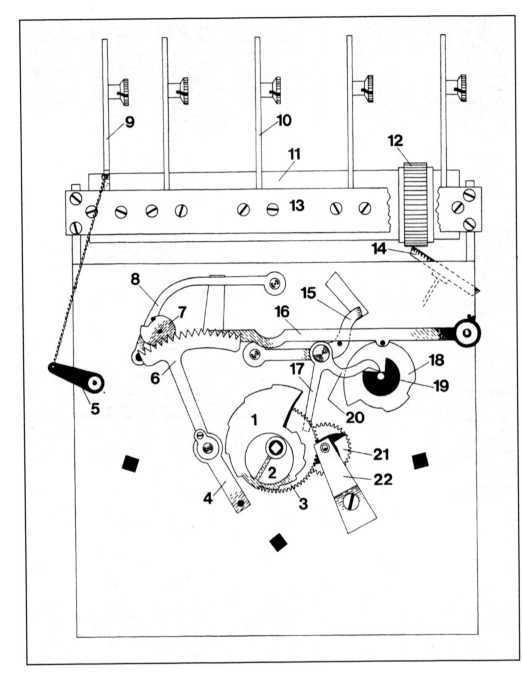

Fig. 54. A Herschede 5-tube movement from the 1920's

Parts list for Figures 54, 55, and 56.

1 snail
2 moon eccentric disk
3 hour wheel
4 rack tail
5 hour strike shaft assy.
6 hour rack
7 gathering pallet
8 hour strike lever
9 hour strike lever
10 chime lever assemblies
11 cylinder
12 cylinder gear
13 hammer bracket
14 crown wheel
15 warning arm C.T.
16 long lever
17 lifting lever
18 locking plate
19 chime correction cam
20 self adjusting lever
21 intermediate wheel
22 intermediate bridge
23 minute wheel
24 hour bridge
25 warning arm S.T.
26 star cam
27 chime third arbor
28 chime lock pin
29 chime locking lever
30 chime locking wheel

Chime

The overall chime train design has some points in common with other Herschede models, but there are quite a few differences. Refer to Figures 54, 55, and 56. The cylinder (11) receives power from the crown wheel assembly (14). Note the angle of the crown wheel shaft. A large bevel gear is located on the second arbor, behind the second wheel. It provides rotation to the small bevel gear on the lower end of the crown wheel shaft. There is no endshake adjustment on the second arbor, because the second shaft bridge assemblies are absent in this design. This is a definite "minus" compared to the more common 5- and 9-tube movements; one of the major bevel gear adjustments is not available. The only

thing you can do is to raise or lower the small bevel gear on the crown wheel shaft.

The intermediate wheel (21) has a star cam (26) instead of pins to raise the lifting lever (17). Chime correction is accomplished on the front of the movement. The chime correction cam (19) has one notch for the end of the third quarter chime. Figure 54 shows this position, with the self adjusting lever (20) seated at the cam. The lifting lever moves to the left, making it impossible for the three short points of the star cam to reach it at all. The long hour point reaches it to unlock the train at the hour. During the rest of the chime sequences, the self adjusting lever rides on the chime correction cam.

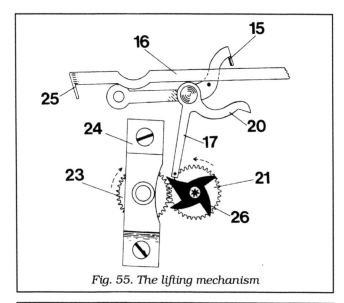

Fig. 55. The lifting mechanism

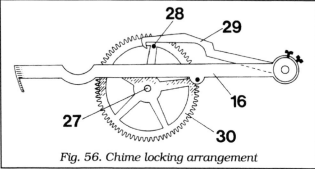

Fig. 56. Chime locking arrangement

This action keeps the lifting lever more to the right, where all points of the star cam can reach it.

Locking action is shown in Figure 56. The long lever (16) is placed across the front of the movement. It is held on the same arbor which carries the chime locking lever. The locking lever (29), lock pin (28), and locking wheel (30) are behind the front plate. The lock pin and wheel rotate counterclockwise and are shown in Figure 56 just before the point of locking. The pin on the long lever rides the locking plate (Figure 54, part#18). Each time the pin drops into a slot, the chime train locks on the next pass of the locking wheel. The locking plate has only three slots, instead of the four slots you would expect. One is a double width slot for the division between the first and second quarters.

Strike

The locking plate has an increased diameter at the hour segment. This lifts up the long lever to release the hour rack hook (8). The train locks on the gathering pallet (7) which has a sharp corner to catch a pin on the rack hook.

To operate the hour strike lever (9), the hour strike shaft assembly (5) pulls on a chain connecting the two parts. The pin wheel (second wheel) provides the lifting action.

Notes on Assembling the Movement

Assembling the movement is relatively easy if you keep a few things in mind. First, remember to install the strike hammer arbor between the plates before you start. It cannot go in later. Next, look at the chime side of the movement. The crown wheel bridge, which supports the crown wheel shaft assembly, is screwed to the back plate. This can be done before or after the plates are put together. But if you studied the movement before you took it apart, you might remember that the chime locking lever (29) goes between the bridge and the crown wheel shaft. Preassembling these parts is difficult. I tried it that way the first time around. That is, I assembled the chime locking lever onto its shaft and loaded them into the front plate with all the other wheel and arbors. Then I proceeded to add the back plate. The crown wheel bridge got in the way. A much easier procedure is to leave out the locking lever and arbor. After the plates are together, slip the lever in place by itself, then insert the arbor through the back plate.

Once the plates are together, proceed with the strike adjustments. The gathering pallet fits on a squared arbor, so you can see if one of the four possible positions allows the hammer tail to be at rest when the strike train is locked. Probably, you'll have to separate the plates enough to change the relationship of the pin wheel with the pinion above. Set the warning pin and warning arm S.T. (25) for a half turn of warning.

Chime adjustments are next. Install all the chime parts. Set the chime lock pin (28) against the chime locking lever (29), and the pin on the long lever (16) in a slot in the locking plate (18). Tighten the two set screws shown in Figure 56 in this position. One screw holds the chime locking lever. It tightens against a flat place on the arbor, so the only adjustment you're looking for there is front to back. You don't want to have the locking lever scrape the wheel or miss the pin. The other set screw is on the long lever. It controls the depth of locking, but if you set it with the parts positioned as described above it should be right. Check for proper locking, unlocking, and chime correction. Set the warning pin to provide about a half revolution of run to the warning arm C.T. (15).

Most of the remaining adjustments are similar to those for other Herschede movements as described in this book.

10

TWO-WEIGHT TUBULAR BELL

Herschede made two-weight tubular bell movements with combined chime and strike trains. These 5-tube movements are fairly common today, so a fairly large number must have been manufactured. There were, however, several versions made.

Figure 57 shows the movement as it appeared in the 1927 catalog. It was captioned "First Quality" and appears to be the better version. The movement had polished plates and brass cable drums. The required time weight is 15 lbs.; the chime-strike weight is 30 lbs.

A lesser quality version of the movement, shown in Figure 58, has the same basic design, but the plates are not polished. The cable drums (Figure 59) are a gray, cast zinc metal and are subject to chipping and breakage, including damage to the integral ratchet teeth. The weights are lighter at 14 lbs. for the time and 20 lbs. for the chime-strike. The pendulum is also lighter than the "standard" Herschede tubular bell pendulum.

There was a less common 9-tube, two-weight movement, shown in Figure 60. The tubes are hung in an offset pattern that maintains the same width as a 5-tube set. The hammer assembly shifts from side to side as Westminster, Whittington, or silent operation is selected.

This chapter will present a description of chime-strike operation for the two-weight chime and strike mechanism. Then a separate assembly and adjustment procedure will be given for the two main 5-tube versions shown in Figures 57 and Figure 58.

CHIME AND STRIKE OPERATION

The two-weight movement plays full Westminster chimes on four tubular bells and strikes the hour on the fifth bell. The order of the tubular bells is: shortest on the left, longest (hour strike) on the far

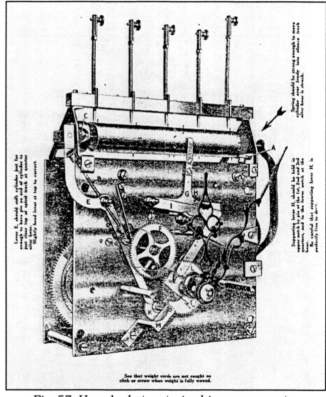

Fig. 57. Herschede two-train chime movement from a Herschede instruction sheet.
Image courtesy Herschede Hall Clock Co.

right. The time main wheel, with maintaining power, is on the left side of the movement. The chime-strike gearing is on the right side.

Figures 61 and 62 were drawn from a movement similar to Figure 57, marked with patent dates of Jan. 12, 1909 and the serial number 8076, which is from 1919 according to the serial number list (Appendix C). Figure 61 shows the main front movement parts for our discussion. Chimes and strike are controlled at the cylinder (6). There is no rack and snail system for either chime or strike. Instead, there is the chime-strike control assembly (4). Lo-

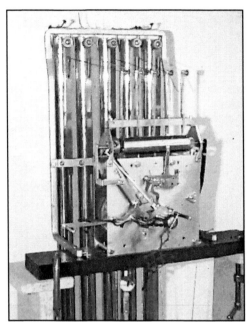

Fig. 58. The "lesser quality" 5-tube, two-weight movement (see description on facing page).

Fig. 59. Cast zinc cable drums are found in the lesser quality movement. They are a dull gray color. An end view is shown above and a side view at the right.

Fig. 60. A 9-tube, two-weight Herschede movement. The dial carries a personalized plate dated 1918. Photo courtesy Robert Webb.

cated to the left of the cylinder, this device measures out the chime notes as a pin on the end of the cylinder traces its way through a long spiral groove. As the pin reaches the end of this groove, the cylinder shifts to the left and the strike begins. The strike counts the correct hour, but the train actually runs for a full count of twelve each hour.

One way to approach this movement is to describe some of the actions taking place each quarter hour. In this way I can cover the operation of the cylinder, chime-strike control assembly, lifting lever, and locking lever.

1st Quarter Chime

Refer to Figure 61. The lifting lever (1) is raised by a pin on the cannon pinion. (There are four of these pins, one for each quarter.) The detent (2) now comes into play. A pin on the underside of the lifting lever (shown dotted) raises the detent, which in turn lifts the locking lever (3). This action accomplishes two things. First, the chime-strike train goes to the warning position. Second, the detent is lifted high enough so the pin on the locking lever is caught in the second notch (right side) of the detent. The detent holds the lever in this position. Chiming begins as the pin

Fig. 61. Herschede two train chime movement, front movement parts controlling chime and strike functions

1 lifting lever
2 detent
3 locking lever
4 chime-strike control assembly
5 chime hammer lever (4)

6 cylinder
7 strike hammer lever
8 cylinder gear
9 hour lock pin
10 quarter hour lock pins

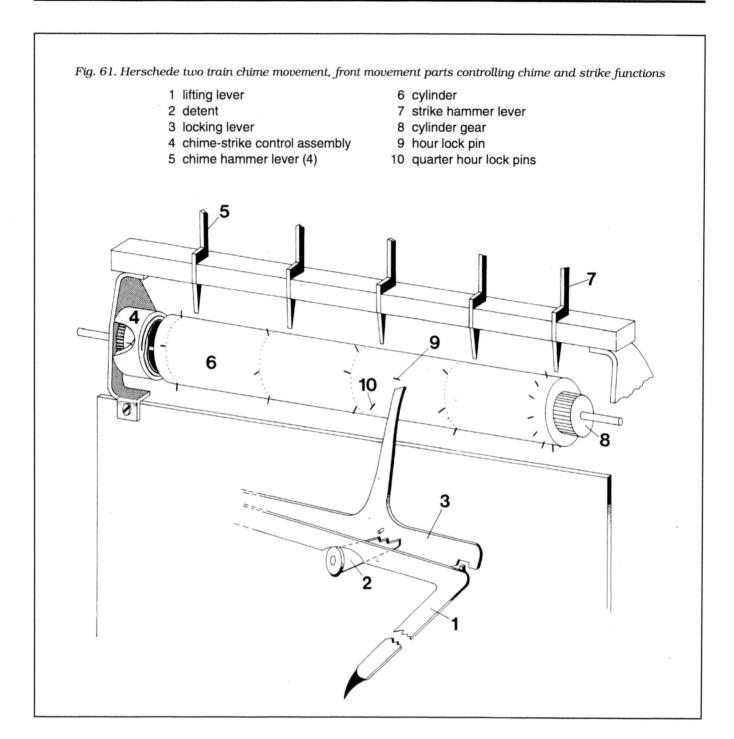

on the cannon pinion releases the lifting and warning levers.

The pin on the end of the cylinder begins to trace through the long spiral groove in the chime-strike control assembly. This is a counting process that keeps the cylinder in the same lateral position through the 40 chime notes played during each hour.

The cylinder must determine, however, where each chime sequence ends. A special row of two pins exists on the cylinder. One of the pins is shown as part 10 in Figure 61. As the cylinder rotates, the

first pin reaches the locking lever. The pin raises the locking lever, and the detent drops. As the pin moves on, the locking lever drops, too, and the train locks.

2nd Quarter Chime

The same actions take place during the second quarter chime. The detent catches the pin on the locking lever in the second notch, keeping the lever lined up the same way as before.

Without a rack or locking plate, the cylinder counts out eight notes for the second quarter. The other pin in row (10) simply comes along and trips the locking lever at the end of the chime sequence.

3rd Quarter Chime

Counting of the third quarter chime is done in the same way, with the first pin in row (10) coming around again to end the chime sequence after twelve notes. The cylinder has been rotating, of course, but it has not changed its side-to-side position during the first three quarters.

The detent has held the locking lever in just the right place so the pin in row (10) would hit it to end each quarter chime.

Hour Chime

One pin on the cannon pinion is closer to the center than the other three, so it provides less lifting action. This is for the hour. Most clocks have a higher lift for the hour, but not the two-weight Herschede. As confirmation, the minute hand goes on only one way (in the better quality version) and now points straight up. The detent is lifted only high enough to catch the pin on the locking lever in the first notch.

The cylinder still hasn't changed its lateral position, but the locking lever has been moved. The detent holds it to the right of its former position. As a result, the locking lever doesn't line up with row (10) anymore. That means that neither of the two pins can trip the lever to stop the train on the hour. The remaining sixteen chime notes, for the hour, are counted out by the pin in the end of the cylinder, tracing its way toward the end of the spiral groove. Remember, this groove has determined the total number of chime notes to be sounded within one hour. When it finally does reach the end, the cylinder shifts left. The four rows of pins for the chime hammers no longer line up with them. Chiming is complete, although the gear train continues to run.

It is interesting that the chime note sequences for the first three quarters are controlled by only two pins in row (10). Locking occurs at the end of the first quarter with the first pin and at the end of the second quarter with the second pin. For the third quarter, the cylinder turns until the first pin locks it again. Then the hour chime proceeds until the cylinder shifts at the end. That brings the cylinder back to the right place to start again for the first quarter chime. The cylinder can then turn only through the four descending notes before locking again. The counting of the first three quarters is controlled by one mechanism and the hour chime by another, and between them they work.

Strike

The cylinder has shifted to line up the row of pins at the far right end of the cylinder with the strike hammer lever (7). Striking begins. From this point, the cylinder always turns one full revolution, no matter what the hour. Whether it is 1 o'clock, 12 o'clock, or anything in between, the train must run out the full count. Here's how it works.

Part of the chime-strike control assembly is a notched hub which signals the end of the strike. The pin in the end of the cylinder moves around the hub as the clock strikes. When it finds the notch, it dips in. The cylinder moves to the extreme left, and the pins move out of line with the strike hammer lever. The gear train continues to run, although there is no further chiming or striking going on. The hour lock pin (9) finally trips the locking lever to stop the train.

After the hour, a lever moves the cylinder to the right to reset it. The pin in the end of the cylinder goes back to the beginning of the spiral groove. The notched hub advances steadily, and it moves into position for the next hour by the time the hour arrives.

Chime Correction

Automatic correction takes place after the actual hour (minute hand straight up), and before the first quarter. But the correction takes place only if the hour chimes and strike have finished. That is, the pin on the end of the cylinder must have run through to the end of the spiral groove, taking the cylinder to the left. This places pin (9) in position to stop the cylinder at the right place.

After the strike completes itself, the cylinder has assumed the far-left position where no hammers can be lifted. The pins do not line up with the hammer levers. The cylinder will turn one full revolution each time the minute hand passes one of the quarters, with no chimes or strike. But after the hour, the cylinder is automatically reset to the right. The pin at the end of the cylinder moves to the beginning of the spiral groove. As the minute hand reaches the first quarter, the correct four notes of Westminster are sounded.

...text continued next page

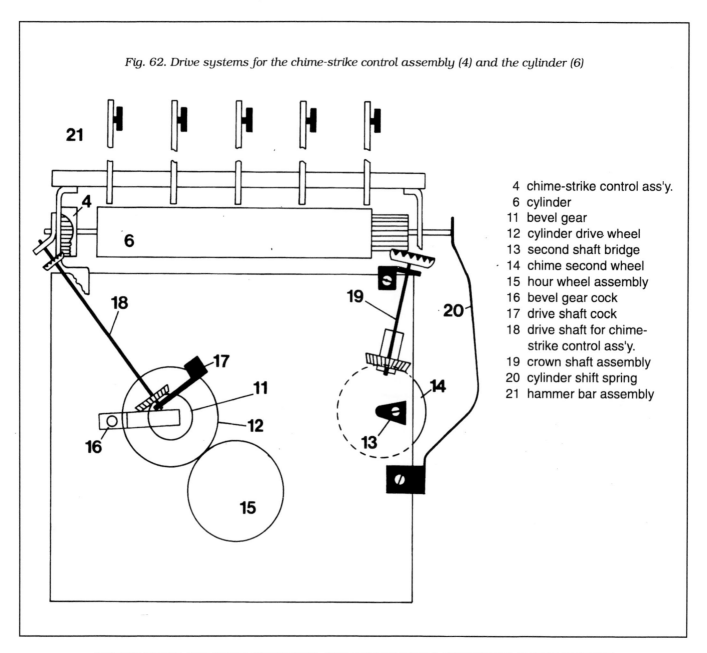

Fig. 62. Drive systems for the chime-strike control assembly (4) and the cylinder (6)

4 chime-strike control ass'y.
6 cylinder
11 bevel gear
12 cylinder drive wheel
13 second shaft bridge
14 chime second wheel
15 hour wheel assembly
16 bevel gear cock
17 drive shaft cock
18 drive shaft for chime-strike control ass'y.
19 crown shaft assembly
20 cylinder shift spring
21 hammer bar assembly

ASSEMBLY OF THE BETTER-QUALITY TWO-WEIGHT MOVEMENT

Disassembly and cleaning chores are necessary, as always. At least, you should be pleasantly surprised by the assembly task which follows. It really isn't that difficult. With large wheels and almost 2-1/2 inches between the front and back plates, you may find it relatively easy to get things together again. Assemble all the arbors into the front plate before adding the back plate.

Assembly

There are five time train arbors, including the center arbor and the pallet arbor. The pallets can go in much later, of course. But don't forget the maintaining shaft assembly (the lever for the maintaining power). Even more important, look for the bevel gear (11). This gear and its long arbor must be fitted into the front plate before the time main wheel is installed. The front pivot support for this arbor is the front plate. The rear support is the bevel gear cock (16), but it isn't necessary to install the cock at this point. If you forget to fit the arbor through the front plate, you must take the entire movement apart again to install it.

The chime-strike train consists of five wheels and arbors, including the fan shaft assembly (fly). The fan shaft assembly can go in later. The crown shaft assembly (19) and its supporting cocks can just as well go in before the plates go together, because these parts do not get in the way. Take a moment to look at the second shaft bridge (13) as you install it. My

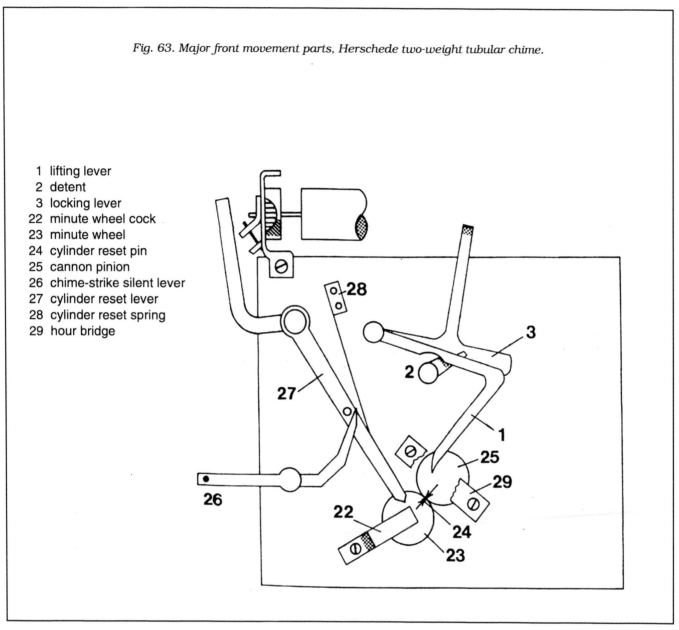

Fig. 63. Major front movement parts, Herschede two-weight tubular chime.

  1 lifting lever
  2 detent
  3 locking lever
 22 minute wheel cock
 23 minute wheel
 24 cylinder reset pin
 25 cannon pinion
 26 chime-strike silent lever
 27 cylinder reset lever
 28 cylinder reset spring
 29 hour bridge

example had one of these on the front pivot hole for chime second wheel (14), but not for the rear pivot hole. In any case, mine has a clearance problem with its mounting screw. This particular screw is the same diameter and thread as all the other pillar and bridge screws on the movement, but is filed shorter. A full length screw hits the chime second wheel, stopping it. I have no way of knowing whether this oddity applies to our example alone, or to other similar Herschede models.

With the plates together, refer to Figure 62 and add the following parts: locking lever (3); detent (2); and lifting lever (1). Use new taper pins. Add the fan shaft assembly and fan bridge. Continue by adding the drive shaft cock (17). Install the bevel gear cock (16). This secures the bevel gear (11) which you had to install before putting the plates together.

Refer to Figure 63 and add the cannon pinion (25). Now install the minute wheel (23) and minute wheel cock (22), Arrows on these two wheels indicate the point they should mesh. The important point is really the cylinder reset pin (24) on the minute wheel. It must reset the cylinder to the right, after the hour and before the first quarter chime. The pins on the cannon pinion establish the four quarters, so this precise meshing of the wheels will assure that the reset occurs at the right time.

Next, install the cylinder reset lever (27). After each hour, the reset pin (24) moves this lever to reset the cylinder to the far right position. The cylinder reset spring (28) holds the lever in position until the next hour passes. The chime-strike silent lever (26) can be added now. When the lever is pulled down, it contacts a pin on the cylinder reset lever, moving it away from the reset pin. Without the reset action, the cylinder just runs in "neutral" and doesn't lift any hammers.

Locate the required parts in Figures 61, 62, or 63. Proceed next to install the left cylinder bracket. The chime-strike control assembly (4) is screwed to this bracket. Before tightening the bracket in place, put in the chime-strike drive shaft (18). The shaft's upper support is on the bracket; the lower support is the cock (17) we installed earlier. The cock is adjustable to some degree, even though it is positioned by two pins. On our example I had to loosen the screw and move the cock slightly to obtain endshake for the drive shaft. After tightening, recheck. A tight shaft could bring things to a halt.

If you haven't done so before, install the crown shaft assembly (19) with upper and lower cocks. Add the cylinder (6) and the right-side cylinder bracket. The cylinder shift spring (20) is next. Check for adequate tension, because the spring must push the cylinder without hesitation. I found that the clock was striking extra notes because the spring wasn't strong enough to move the cylinder after the last note. The cylinder always runs for a full 12 count each hour. In conjunction with the chime-strike control assembly, the cylinder shift spring must move the cylinder out of line with the strike hammer lever. It has less than a second to do this before an extra note is struck The hammer bar assembly (21) can now be added across the cylinder brackets.

Adjustment

The assembly procedure is nearly complete. I suggest the following steps, required to synchronize the chime-strike mechanism with the hands.

1. Put on the minute hand (only one way). Turn hand straight up to 12, then remove it.
2. Install the cylinder drive wheel and pin it in place on the arbor.
3. Turn the cylinder drive wheel by hand, which moves the drive shaft and the chime-strike control assembly. Stop when the numeral 12 stamped on the brass gear in the control assembly lines up with the line on the hub. This is the 12 o'clock position.
4. Now install the hour hand on the hour wheel assembly. Put the assembly in place with the hour hand pointing up to 12. The hour hand agrees with the control assembly.

The rest of the movement is self adjusting. The movement is easy to assemble because the chime and strike control depends on the cylinder and the chime-strike control assembly. The warning action is preset and not adjustable. The locking action

takes place between the locking lever (part #3 on Figures 61 and 63) and the locking wheel (not illustrated) which has four locking pins. No adjustment is required there either.

Every good thing has its downside, and I suppose this movement has its flaws. The grooves in the chime-strike control hub are likely to wear out the brass pin on the cylinder. Dirt and wear create a sticky action of the cylinder, causing counting errors. Polish the cylinder pivots to remove wear, and apply a good quality clock grease to the grooves in the control hub. Finish the job with a few weeks of testing.

THE LESSER-QUALITY MOVEMENT

This movement is also relatively easy to assemble and set up because all the controlling parts are easily reached on the outside of the movement. Here is a suggested assembly and adjustment procedure. It is only slightly different from the previous section describing the steps to follow for the better-quality movement.

Assemble all the arbors between the clock plates. When installing the motion work, mesh the cannon pinion and minute wheel to line up the arrows. These gears are shown in Figure 64. Add all the rest of the front movement parts. Do not install the weights.

Fig. 64. Mesh the cannon pinion and minute wheel to line up the arrows.

Put on the minute hand and turn it through several quarter hours. Notice the levers in the center of Figure 65. The detent is the notched lever set at a slight angle. Each time a chiming point is approached, the detent will be lifted so that it locks on either the first or second notch. For the first three quarters, it will be lifted to the second notch. For the hour only, it will be lifted to the first notch. Once the hour location is determined in this way, install the hour hand first, then install the minute hand pointing straight up.

Next locate the long, diagonal steel arbor which connects the chime-strike control assembly with the hour wheel assembly. The purpose of this gearing

Fig. 65. The vertical locking lever at the top center of the photo shuts off the gear train at the end of each chime or strike sequence.

Fig. 67. At the end of the hour chime, the cylinder must shift quickly to the left to line up the row of 12 strike pins with the hammer.

is to move the chime-strike control assembly to the next hour. Loosen the set screw on the crown gear at the lower end of the arbor and move it out of mesh with the hour wheel assembly. Turn the gear at the top end of the arbor by hand. This will turn the chime-strike control assembly through the hours. Stop when the notch lines up with the line indicating the numeral 12 on the hub of the assembly (Fig. 66). This is the 12 o'clock position. Mesh the crown gear with the hour wheel assembly and tighten the set screw.

The hour hand can be turned independently of

the rest of the motion work. Set it at 12. With both hands pointing straight up, the clock is set up and adjusted for correct chiming.

Test the movement for correct chime and strike. Check to be sure the cylinder shifts smartly to the left immediately following the hour chime. Figure 67 shows the strike hammer lever lined up with the row of strike pins on the cylinder. If the shift is sluggish, the hour strike count will be wrong. If necessary, increase the tension on the large steel spring on the right side of the movement.

Fig. 66. At the arrow, line up the notch with the line on the hub, indicating 12 o'clock.

11

CHIMES

Chapter 4 covered Herschede chimes in terms of assembling and adjusting the movement, without regard for the chime melodies as such. If you put the movement together correctly, you usually don't have to be concerned about the musical aspects; the clock corrects any chiming errors automatically.

You could think of the Herschede tubular bell clock as a large musical instrument that tells time. Still, you don't have to be a musician to work on a Herschede. It's just that repairs will be easier for you if you have a basic familiarity with the notes. Figure 68 is a table of the 9-tube chime sequences for all three chime melodies. Figure 69 presents the same data for a 5-tube clock playing Westminster chimes. Figure 70 is a Herschede illustration showing customers the written chime notes and some background on each melody.

Don't forget to review the mechanical aspects of chiming. To have correct chimes, you require a movement that works properly; the chime and strike mechanisms must lock, unlock, and count dependably. The self correcting assembly must function. The cylinder must be installed the right way, and the hammer bar spring needs to be strong enough. All the chime tubes must hang in the right locations. With this firm foundation, you will do fine if you can distinguish one chime from another.

WESTMINSTER

Westminster is, without a doubt, the most easily recognized, popular chime–the chime of "Big Ben". It is built upon a four note sequence. Refer to Figure 68 to learn the pattern. Westminster is chimed only on tubes 3, 4, 5, and 8 in the 9-tube clock. The 5-tube clock makes use of all four chime tubes, numbered 1, 2, 3, and 4 (Figure 69). There are five

different note patterns in the Westminster chime, utilizing only these same four notes. At the first quarter, the descending four notes are played alone (one sequence). The half hour chime is made up of eight notes (two sequences of four notes each). At the third quarter, the clock chimes twelve notes (three sequences), and you should note that the last four notes played are the same ones played at the first quarter. The hour chime consists of sixteen notes (four sequences). Some clock owners say Westminster is the "slowest" chime, and they want to know why this is so. It is simply because this chime has the fewest notes to play per revolution of the cylinder. Forty notes are played in the course of an hour. Westminster corresponds to the far right-hand position of the 9-tube cylinder.

CANTERBURY

When the 9-tube cylinder is shifted to the center position, the clock plays the Canterbury chime. Each note sequence is made up of six notes instead of four. This means that although the same number of sequences are played, each has more notes. Each hour, 60 notes are played. Tubes 3, 4, 5, 6, 7, and 8 are utilized. Figure 68 presents the pattern of notes played. Canterbury sounds "faster" than Westminster because more notes are played in each sequence. The Canterbury chime is less familiar to repairers. A pattern may be observed, however. The first quarter chimes, consisting of six notes, are exactly the same as the last six notes of the third quarter chime. Identify tubes 3-5-7-4-6-8 (number 8 is the tube with the lowest note, hanging on the far left). If you look for this sequence of hammer strokes, you can be sure the chime is correct even if you do not know the the melody. Upon checking a recent model 9-tube Herschede clock, I found a

Quarter		WESTMINSTER	CANTERBURY	WHITTINGTON
Quarter	1	* 3 4 5 8	* 3 5 7 4 6 8	* 1 2 3 4 5 6 7 8
	2	5 3 4 8	7 5 3 6 8 4	1 7 2 6 3 5 4 8
		5 4 3 5	5 3 7 4 8 6	1 3 5 7 2 4 6 8
	3	3 5 4 8	8 3 5 7 6 4	1 2 5 6 3 4 7 8
		8 4 3 5	5 3 7 4 8 6	7 5 3 1 2 4 6 8
		* 3 4 5 8	* 3 5 7 4 6 8	* 1 2 3 4 5 6 7 8
	4	5 3 4 8	7 5 3 6 8 4	1 7 2 6 3 5 4 8
		5 4 3 5	5 3 7 4 8 6	1 3 5 7 2 4 6 8
		3 5 4 8	8 3 5 7 6 4	1 2 5 6 3 4 7 8
		8 4 3 5	5 3 7 4 8 6	7 5 3 1 2 4 6 8

The numbers shown in the table above are chime tube numbers, listed in the order they chime in each sequence of notes.

Tube number 1, the high note, hangs on the far right and is the shortest. Number 8, the longest chime tube, hangs on the far left. The hour tube, number 9, hangs on the right, in front of the others.

* On all the chimes, the first quarter chime sequence repeats as the conclusion of the third quarter chime. Although all the note sequences repeat, this one is easy to recognize, especially in Westminster, with four descending notes; and Whittington, with eight descending notes.

Fig. 68. The 9-tube chime sequences

difference between the way the clock chimed Canterbury and the written musical notes as published by Herschede. The clock chimed the last six notes of the third quarter chime as shown above: 3-5-7-4-6-8. The written musical notes show this sequence to be 3-5-7-6-4-8. This variance, while interesting, does not affect you as a repairer. It is the locations of the pins on the cylinder which determine the pattern of notes.

WHITTINGTON

If you remove the 9-tube clock dial, the clock will be set to chime Whittington. The leftmost position of the cylinder corresponds to this chime, and the hammer bar spring pushes the cylinder in this direction. Whittington has eight notes per sequence; it makes use of all eight chime tubes (number 9 is the hour tube). The first quarter is easy to recognize, with eight descending notes played down the scale. The hammers will rise and fall from right to left, in order, 1-2-3-4-5-6-7-8. Figure 68 shows the

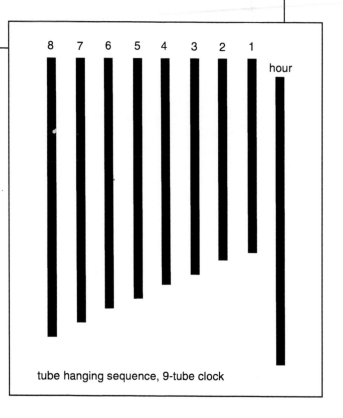

tube hanging sequence, 9-tube clock

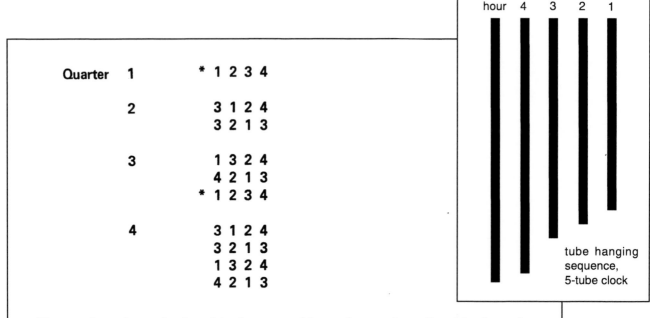

Quarter	1	* 1 2 3 4
	2	3 1 2 4
		3 2 1 3
	3	1 3 2 4
		4 2 1 3
		* 1 2 3 4
	4	3 1 2 4
		3 2 1 3
		1 3 2 4
		4 2 1 3

The numbers shown in the table above are chime tube numbers, listed in the order they chime in each sequence of notes.

Tube number 1, the high note, hangs on the right. The longest tube hangs on the left. In recent 5-tube clocks, this tube is number 5, the hour tube. Some older models place number 4 on the far left, and number 5 on the right, in front of the other tubes.

* The first quarter chime, four descending notes, repeats as the conclusion of the third quarter chime. Each of the five different note sequences repeats, but this one is easiest to recognize as you adjust a 5-tube clock.

Fig. 69. The 5-tube chime sequences

pattern of the notes. As you adjust the clock, always check it on Whittington. All the notes are played, giving you a good opportunity to hear any difference in loudness and locate incorrect hammer clearances. Whittington is also the best check on the cylinder adjustment. This is because more notes are played per revolution of the cylinder, which means there is less tolerance for error. If the last hammer is going to "hang" instead of falling to strike the tube, it is most likely to show up on Whittington.

A WORD ABOUT CHIME MELODIES

Always have your repair customer listen to the chimes when you complete the work on the clock. Try all three chimes on the 9-tube clock. If any notes sound too loud or soft, adjust them to suit the owner. Then ask which chime the owner would like selected. This is most likely Westminster, the most popular. Some customers change the chimes often, but most of them do not.

Fig. 70 (opposite page). This Herschede page presented customers with some background on the three chimes used in the 9-tube clock. The musical notes for each chime are shown under the descriptions.

Herschede

Chiming Floor Clocks

REFLECTING A TIME-HONORED HERITAGE of master craftsmanship, the Herschede floor clock has long stood in the American home as a symbol of taste and refinement. In the serene charm of its mellow chiming, you will find a compelling yet unobtrusive sentinel as it marks the passing hours. And, as it continues to note the hour, year after year, you will know the deep satisfaction that comes with the possession of an instrument of rare beauty and of mechanical perfection.

OVER FOUR GENERATIONS of master craftsmen have created in Herschede clocks that intrinsic excellence recognized universally by those who know quality in precision movements. Moreover, these exquisite floor clocks represent the unique combination of technical skill and traditional design; the fine cabinetwork and delicate dial engraving reflect the inspiration of revered classical standards. No matter which of the two famous chiming movements you choose, you will come to realize with thrilling pride that you have established a priceless heirloom for the generations of the future.

WESTMINSTER CHIMES

Traditionally rung from the Victoria Clock of the House of Parliament, these beloved chimes are correctly reproduced on four tubular bells. The original melodic inspiration is to be found in the fifth bar of Handel's magnificent symphony, "I Know That My Redeemer Liveth."

Lord, through this hour
Be thou our guide
So, by thy power
No foot shall slide.

WESTMINSTER

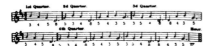

CANTERBURY CHIMES

This beautiful carillon was inspired by the great Cathedral of Canterbury. Played on six tubular bells, it has long been cherished among the finest musical expressions of its kind.

CANTERBURY

"I can pass this way but once.
Any good, therefore, that I can do, let me know it now.
Let me not neglect it or defer it.
For I shall not pass this way again."

"Turn again, Whittington,
"Lord Mayor of Londontown."

WHITTINGTON CHIMES

Accurately rendered by eight bells, this old English melody first belonged to the Church of St. Maryle Bow, from whose chimes the boy Dick Whittington is said to have received the calling to return and become "Lord Mayor of Londontown."

WHITTINGTON

12

TROUBLESHOOTING

Every repairer needs to have good troubleshooting skills. Troubleshooting requires the ability to observe a defect keenly, and to make just the right correction to fix the problem. You may be called to a customer's home to do this kind of work, or you may troubleshoot problems on movements you overhaul in your shop. Even if you have done a good job of cleaning and assembling a movement, defects will show up on the test stand. Of course, this is just where you want them to show–not in the customer's home after delivery!

Like any other clock, the Herschede tubular bell movement has its own characteristics and common ailments. No listing of this kind could ever be complete, but many of the problems you could run into are shown here. The material is organized into a section for each gear train, followed by a question and answer section.

TIME TRAIN
The clock stops randomly and will not continue to run for any length of time.
- Movement dirty; cleaning required.
- Entire movement needs oiling.
- Escapement is dry of oil. Take off the time train weight, remove the anchor, brush clean the pivots and pallet faces, peg out the holes, put the anchor back in, and oil the escapement.
- The clock could be slightly out of beat. Set the beat, but check to make sure the beat adjusting disc on the pendulum is not loose. If necessary, tighten the nut to prevent the clock from getting out of beat again. Refer to Chapter 13 for beat setting instructions.
- The pin could be out of the crutch slot, pressing on the side of the crutch instead. Place

the pin back in the slot.
- The suspension spring might be kinked; replace it. The spring is shown in Figure 71.

The clock stopped 4-6 hours after set-up.
- You left the wooden pulley blocks on! Take them off and restart the clock.
- Check the beat and see if the cabinet is still level.

The clock stops about every 12 hours.
- Jammed moon dial. See Chapter 8.
- Hands catching on some part of the dial or on each other.

The clock stops frequently, and always several minutes before a chime point.
- The chimes may be stalled. See the next section, on the chime train.
- May indicate general lack of oil or dirty pivot holes. The extra load of the chime and strike levers can stall the clock if it is in marginal condition.

The clock will not run more than a minute before it "catches" and stops abruptly.
- Check for a bent escape wheel tooth. Repair or replace the wheel.
- See if the second hand scrapes the dial.

Although the clock continues to run, it seems to suddenly lose time. The chime and strike weights are lagging behind the center weight.
- Check the minute hand tension. If it seems weak, remove the minute tube and check the crimp which provides the hand tension. Make

Fig. 71. Herschede suspension spring (bottom) and pendulum spring shaft assembly, part 2675 (top)

this notch *slightly* deeper, reinstall the minute tube, and try the minute hand again. The hand should offer firm resistance without being too hard to turn.

CHIME TRAIN
Chimes are slow.
• Unbalanced chime train fan shaft assembly (fly). The blades are adjustable.
• Lack of oil, especially higher in the chime train.
• Dirt packed in pivot holes or pinions.
• Excessive hammer tension. Check the row of thumbscrews under the hammer springs on the tube rack. Try backing off on them if they appear to have been turned down.

Chimes do not work at all.
• The heavy chime weight may be hanging on one of the other trains. The weights are marked on the bottom; the chime weight belongs on the right.
• The eccentric adjustment on the fan shaft front pivot may have been turned. Check for screwdriver marks on the slot, but do not make an adjustment unless you are sure this is the problem.
• Cable crossed over itself or jammed in front of the second wheel.
• Second shaft bridges adjusted tightly, leaving no endshake for the second arbor.
• Warning pin jammed on the warning lever. See Chapter 4.

Chimes continuously.
• Gathering pallet worn, damaged, ill-fitting or

missing entirely. Replace it.
• Damaged quarter rack; replace or repair.

Will not resume chiming after being silenced and then turned back to chime.
• Go to Chapter 7 for this one.

Does not play the right sequences of notes.
• The self adjusting assembly may not work. See Chapter 4.
• The cylinder is installed incorrectly; install as explained in Chapter 4.
• Is the chime indicator hand put on the wrong way?
• Tubes hanging from the wrong pegs on the rack. Chapter 11 gives directions.

Clock plays strange, discordant sounds.
• Chime indicator hand put on the wrong way; hand seems to point to the middle of the chime name on the indicator circle, but actually points halfway between two chimes. This leaves the cylinder out of position.
• Two or more tubes are touching each other at the bottom.
• The 3 chime shifting cam, behind the indicator hand, is staked onto the shaft out of position. The chimes do not play correctly because the cylinder is pushed to one side or the other. Remove the 3 chime shifting cam assembly, analyze the problem, and move the cam before restaking it.

Chime weight always becomes lower than the other weights, even though they are wound evenly to start.
• The gathering pallet mislocks frequently, causing the clock to chime an extra sequence or two and then run additional chimes on the next hour to self-adjust. Replace the gathering pallet.
• Poor contact surface between the rack and rack hook causes the hook to bounce back up. Look for burrs or rough spots, and watch the locking action closely until you observe the problem actually happening. Polish the parts to a smooth finish, but do not do any drastic filing.
• On older movements, the rack spring is a flat type instead of a coil spring. If the spring tension is too great, the rack moves so quickly that the rack hook does not seat itself between rack teeth. Reduce the spring pressure.

Chime weight is observed to be higher than the others, although it is wound even with them at

the beginning of the winding period.

• The chimes are stalling frequently.

• Check for all the causes above, under "chimes slow" and "chimes do not work at all".

The chimes begin early or late.

• See Chapter 5 for the adjustment procedure.

• If chiming begins almost 10 minutes early, see if the warning pin is sheared off.

STRIKE TRAIN

The strike is slow.

• Check for unbalanced strike train fan shaft assembly (fly). The blades are adjustable.

• The blades are balanced, but they are pulled too far out from the arbor. Fold them inward to form a smaller diameter fan. This will increase the strike train speed.

• Dirt.

• Lack of oil, especially in the higher pivot holes.

• Excessive tension on the strike hammer. Check the thumbscrew at the base of the hammer spring, on the tube rack. Also look at the hammer cord. If it is taut, the hammer arm may be pulling the hammer too far back against the hammer spring tension. Loosen the cord, checking for correct sound from the tubular bell.

No strike.

• Check for jamming of the warning lever on the warning pin, caused by the warning wheel being out of position. Go to Chapter 4.

• Cable crossed over itself.

• The eccentric adjustment on the front pivot of the fan shaft arbor may have been turned. Check for screwdriver marks on the slot. Do not make an adjustment unless one is needed.

Continuous striking.

• Gathering pallet damaged, worn, or missing. Replace it.

• Damaged hour rack; repair or replace.

Counts all hours incorrectly, and is off by the same error each time.

• Check the hour hand bushing. It may be staked on in the wrong position. A tool such as the one in Figure 72 permits the bushing to be moved with relation to the hand.

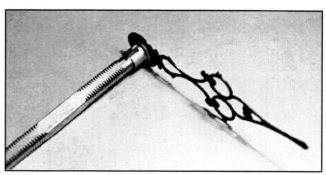

Fig. 72. A homemade adjusting tool is shown inserted into the hour hand bushing. It is a steel shaft with a dull steel "blade" fixed to it. The blade, about .035" thick, fits the slot in the hour hand. The shaft of the tool is gripped with a wrench or could be fitted with a handle. It is held as the hour hand is moved against its bushing.

Counts correctly up to a certain hour, say 8 o'clock, then strikes 8 at 9, 8 at 10, and so on.

• This is an hour silent lever problem. Go to Chapter 6 on the hour silent lever.

Rapid clicking sounds as the clock strikes

• See Chapter 7 on the chime silent lever.

Strike weight lags behind the other two.

• The strike train stalls frequently. Check causes such as dirt, lack of oil, unbalanced fan shaft assembly, and excessive hammer tension.

The strike weight drops faster than the other two weights.

• Check for a gathering pallet which occasionally fails to lock. Replace it.

• Look at poor contact surfaces between the rack hook and rack, which cause the rack hook to bounce back up instead of locking. Polish out rough spots and burrs, but do not alter the basic shape of the parts.

• On older movements, the rack spring is a flat type instead of a coil spring. Excessive spring tension can cause the rack to move so rapidly that the hook cannot seat itself between the teeth.

QUESTIONS AND ANSWERS

Q. What do you consider a proper maintenance schedule (oiling or complete disassembly and cleaning) for a Herschede tubular bell clock?

A. Herschede clocks should not be left to run for many years without maintenance. A call for service may come after 3 to 4 years anyway, because of slow chiming or stopping. In some cases a Herschede may need to be disassembled for cleaning and repair of worn pivot holes after 5 to 6 years.

If you were planning to set up a schedule with your customer for planned maintenance, I would suggest 3 years for oiling, with the idea that after 6 to 9 years the movement will need to be cleaned in your shop, along with any other necessary repairs. Cables should be checked for fraying each time the clock is serviced.

Q. A 9-tube Herschede is giving me problems by chiming on and on. The cause is the self adjusting lever assembly, which raises and holds the quarter rack hook. It never lets go, so the clock cannot stop chiming. How can it be repaired?

A. The photo (Figure 73) shows the self adjusting lever assembly holding the quarter rack hook in the raised position. It does this so the clock will continue to chime while waiting for the long pin on

Fig. 73. The self adjusting lever assembly is shown holding up the quarter rack hook.

the cylinder to come around and trip the self adjusting assembly and the rack hook. This signals the beginning of the hour chime, as the clock can now begin counting the notes.

The self adjusting lever is stamped from thin, soft steel in the later movements, and it is easily bent or damaged.

On your movement the long pin is apparently not raising the self adjusting lever enough so that it, in turn, releases the quarter rack hook. Bend

the self adjusting lever so that more lifting action occurs.

Q. My 9-tube movement would run for about a day and then stop. I found that a hammer on the Canterbury chime was hanging up. I corrected the cylinder adjustment and the problem went away. However, I don't know why this adjustment was necessary in the first place. Initially, I had set up the chimes on the first quarter of Westminster. That should have taken care of it.

A. With any clock that has more than one chime melody, set up the chimes on the melody with the most notes. That is, set it up on Whittington, not Westminster. With the most notes to play, the "set-up" chime requires the finest adjustment of the pin barrel. If you avoid having hammers stopping and starting under load with this chime, the others are almost sure to be correct. In your case, Canterbury has more notes than Westminster, so it guided you to adjust the chimes better. A hammer that "hangs up" can stall some movements.

Q. Recently I was called to repair a 9-tube Herschede clock. The movement was very dirty, and the customer admitted that it had been sprayed with WD-40®.

I removed the movement, took it to the shop, and prepared it for cleaning. Upon closer inspection, I found the #2632 chime train gathering shaft required new bushings in both plates. Following the reassembly, I ran into a problem. The clock will always chime when the hands are turned, but when the clock is allowed to run on its own, the chimes often fail to move to the warning position. A simple touch to the gathering pallet or any other part of the chime train will cause the movement to "warn".

I have carefully inspected everything. I have run the chimes with the crown wheel #2152 (and everything above it) removed. I have cleaned the movement twice. The only part that is slightly worn is the rear pivot hole for the chime warning arbor.

A. You have already eliminated the hammer assembly and the crown shaft assembly from the problem by disconnecting them. Having narrowed the search to this extent, you should double check that all chime train arbors have endshake and that none of the pivots fails to go all the way through its hole. A pivot can tunnel into a hole and cancel out the endshake, even though the pivot and hole have a normal appearance from outside the plates.

Next I would check the fan shaft assembly (the fly). Check the pinion with a loupe to see whether the leaves are rough or clogged with dirt. Then look at the position of the movable fan blades. Fold them in or out to balance them. You can try folding both blades inward to adjust the chime train for slightly faster running and less resistance to starting. Also consider an adjustment to the eccentric fan shaft bushing in the front plate. Telltale marks around the slot may indicate that someone has turned the eccentric bushing and introduced an incorrect depth to the fly pinion. Use a correctly sized screwdriver with parallel faces on the blade to twist the eccentric bushing and change the depthing of the fan (fly) pinion with the wheel below it. On the principle of leaving well enough alone, I would suggest that if there is no sign that the bushing has ever been moved, it does not need to be moved now.

Q. A Herschede 5-tube grandfather clock movement with 1920's patent dates is now in my shop. Does this movement require straight (rather than offset) gathering pallets for both the chime and the strike? In trying to replace the chime pallet, I obtained an offset type pallet which does not have the correct hole size.

A. For this older movement you can use part 22106 (Figure 74), currently the strike gathering pallet, for both chime and strike. Note that when you replace any Herschede gathering pallet, it may be necessary to file and polish the ends to work

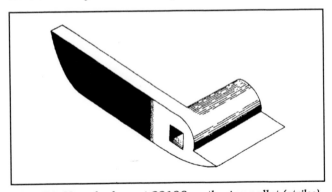

Fig. 74. Herschede part 22106, gathering pallet (strike)

properly in the particular movement on your bench. The long end may be too long to clear the pin during the last revolution before locking, and the shorter pallet end may engage the rack too deeply.

As for the problem with the hole size, Herschede owner Randy Thatcher suggests that you adapt your older movement to use the modern replacement part by polishing a few thousandths of an inch off the flats on the gathering shaft. The change in hole size resulted from Herschede's 1980 retooling effort. It is always necessary to fit gathering pallets individually to the movement. Randy Thatcher also suggests that you check for a worn pivot hole at the gathering pallet location and make sure the lifting lever extension is not worn or damaged. These other problems can make it impossible to successfully replace

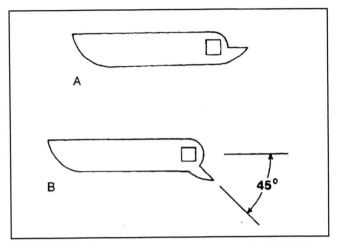

Fig. 75. A comparison of the gathering pallet (strike) 22106, shown as **A**, and gathering pallet offset (chime) 22109, shown as **B**.

a gathering pallet.

The offset chime gathering pallet, part 22109, was introduced with the 1980 retooling. The offset pallet increases the interval between the end of the hour chime and the beginning of the strike. It does this by causing the chime to end earlier than before. This improves the performance of some Herschede movements which sound the first strike note almost on top of the last chime note at the hour. Figure 75 compares the two styles of Herschede gathering pallet.

I repaired a 5-tube movement with 1914-1921 patent dates. The chime gathering pallet found on the clock was an offset type. This would indicate that it is at least possible to use the offset pallet 22109 on the chime side of an older movement, especially if there is a need to lengthen the interval between chime and strike.

Q. What lubricant should be used on the winding assemblies and on the large pivots? What oil is best for the other pivots?

A. Disassemble the winding drums from the main wheels for cleaning. Before putting the parts back together, apply a light coating of clock grease on the contact surfaces. Some clockmakers use a light grease on the heavily loaded winding arbor pivots, preferring a lubricant that is heavier than clock oil; others use clock oil. For the rest of the pivots, use your regular clock oil.

Q. I am repairing a two-weight Herschede tubular bell movement. It needs a lot of work. The parts that worry me are the cast winding drums, which need teeth replaced, and the chime-strike control assembly at the left top part of the movement, which is worn. Do you have any suggestions?

A. These are specialty clockmaking jobs that I

would refer to a specialist. I suggest you contact James L. Christian III, 750 South Main St., Orrville, OH 44667, phone (330) 682-2206. Mr. Christian has remade the cast ratchet wheels for the two-weight movements. He has also machined replacement chime-strike control assemblies (which he calls scrolls) from castings.

Q. What type of cord should I use to hang the tubular bells? What type of cord is best to connect the hammer levers to the hammer springs?

A. You can use 1/8" braided nylon fishing cord for the tubular bell cords, or you can purchase the cords from Herschede. The hammer cords can be made from 50-lb. test, bait-casting fishing line. It's easier to buy the cords from Herschede as a set of 15 hammer spring strings, part 2386.

Q. Although I have completely cleaned and repaired the movement, this Herschede 9-tube movement chimes sluggishly and sometimes stalls. What have I missed?

A. Two areas relating to the chime hammers need to be checked out. The first of these is the hammer assembly. The second relates to the hammer springs.

You may find that on the hammer assembly the rest position of the hammers has changed to the point that each hammer must be lifted sooner and taken further by the cylinder. This extra strain may be the cause of your problems. Some of the latest movements have a knurled thumbscrew at each hammer lever. Screw this in slightly to reduce the hammer "throw". On many other Herschede movements, the hammer assembly is similar to the sketch in Figure 76. Behind each hammer lever there is a small hole. This cavity contains a leather plug which acts as a bumper or stop for the hammer lever. A brass retaining strip holds the leather plugs in place and acts as a backup for the levers to lean against if the leather pieces deteriorate. As the dotted image shows in Figure 76, a hammer which leans against the brass strip increases the hammer pull.

The cure for this problem is to pull out the steel rod upon which the hammer levers are mounted. Number the levers as you remove them, so they can be put back in the same order. Clean the rod and the hammer levers and set them aside. They should not be oiled, since that may cause sticky operation later on. Remove the brass retaining strip and take out the remains of the leather plugs. Cut new ones with a leather punch, then glue them in place. Now put the hammer assembly back together. It should work more smoothly now.

The second area to check is the set of flat steel hammer springs. Recent movements have thumb-

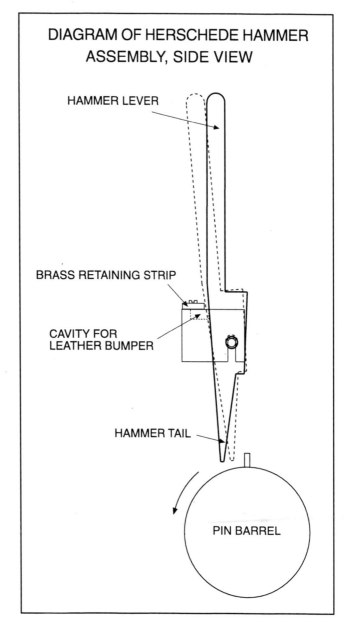

DIAGRAM OF HERSCHEDE HAMMER ASSEMBLY, SIDE VIEW

HAMMER LEVER

BRASS RETAINING STRIP

CAVITY FOR LEATHER BUMPER

HAMMER TAIL

PIN BARREL

Fig. 76. Diagram of a Herschede hammer assembly, viewed from the chime side.

screw adjustments which should be checked. Back off on the tension if any of the screws have been tightened too much. This should reduce the load on the hammers. Older movements do not have an adjustment. Instead, the hammer springs are clamped onto the tube rack with a steel strip screwed in place. If the tube rack has been handled roughly, the hammer springs may have been bent. If they are bent backward, it will take extra force to raise the hammers. It's always possible that the blue steel springs may break if they have to be straightened. In that case they will have to be replaced with new ones. Otherwise, removal of the bends in the hammer springs may return them to the correct "set". It will take less power to raise the hammers.

13

- **HERSCHEDE *OWNER'S MANUAL***
- **AUTHOR'S NOTES ON SETUP**

In this chapter, a Herschede owner's manual on the tubular bell clocks is reproduced and adapted with Herschede's permission. The booklet was originally printed in the 1970's or early 80's and was shipped with new grandfather clocks. Several versions of the grandfather clock manual were published. One version covered tubular bell clocks and chain wound, rod chime clocks. Another version, which is presented here, is a shorter manual covering just the tubular bell clocks.

Please note the gap in illustration numbers from Figure 18 to Figure 25; this resulted when Herschede deleted the chain wound section from this version of the manual without renumbering the illustration and text. Also note that the illustration "Position of tubular bells" on page 15 of the manual had an error in the representation of the tubes in the Model 250 clock. I substituted a corrected illus-

tration from another version of the *Owner's Manual.*

This tubular bell manual contains much useful information to help in setting up one of these clocks, especially for the person who is not trained as a clock repairer. The booklet also has interest for collectors or owners of Herschede tubular bell clocks. The original booklet format was 5-1/2" x 8", saddle stitched (stapled binding), printed on white paper, with a blue-gray ink. This facsimile of the manual includes all the pages in the featured version, except the back cover.

The second part of the chapter presents the author's notes on setting up the tubular bell clocks. The information in this section is based on experience gained in setting up new Herschede tubular bell clocks in the late 1970's and early '80's and servicing movements up to the present time.

OWNER'S MANUAL

cover of Owner's Manual

title page

Owner's Manual

Grandfather Clocks

Tubular Bell (Cable Wound)

**Read this manual in its entirety before
attempting to set up your Herschede clock.**

Herschede Hall Clock Co., Starkville, Mississippi 39759

Contents

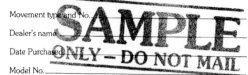

Dear Herschede Clock Owner:

Congratulations!

You now own one of the world's finest chiming floor clocks. Get to know it well — you will be living together a long, long time. Given proper care and maintenance, your Herschede clock will serve you faithfully and accurately for years to come. Before long you will come to think of it not as a mere timepiece, but as a beloved companion that radiates warmth and charm throughout your home.

One of the many features that makes your clock unique is the name Herschede on the dial. This honored and tradition-rich name is the oldest in the American chiming floor clock business. We have been making gold medal winning clocks since 1885. We are, in fact, the only clock maker in America that makes its own tubular bell clock movements. We are clock makers. Not just cabinet makers.

Now, before going further, let's look briefly at the individual parts of your clock.

4

The Case Your case is, above all, a fine piece of furniture — superbly designed and hand-finished. It will become a center of interest in your home. A conversation piece that brings praise and compliments from friends and acquaintances. A symbol of your good taste. Herschede clock designs are rich in tradition, many going back to the 18th Century. With their fine propor-

tions and with the warmth and richness of their coloring and finishes, these clocks not only add magnificence to traditional settings but give modern arrangements a touch of grace and elegance.

The Movement Herschede is the only manufacturer in America that makes its own clock movements. With regular oiling and cleaning, this movement will perform accurately and dependably for years to come.

The Chimes All Herschede clocks play the Westminster chime, the most famous of chime melodies. Based on Handel's "Messiah," "I Know That My Redeemer Liveth," it was made famous in 1859 by London's fabled Big Ben. Besides Westminster, Herschede nine tubular bell clocks play the Whittington and Canterbury chimes. The latter, inspired by the grandeur of Canterbury Cathedral, was composed especially for Herschede.

Whittington

"Turn again, Whittington
Lord Mayor of Londontown."

Canterbury

"I'll pass this way but once
Any good, that I can do,
Let me know it now
Let me not neglect it or defer it
For I shall not pass this way again."

Westminster

"Lord, through this hour
Be Thou our guide
So, by Thy power
No foot shall slide."

The tall and dignified Herschede tubular bell grandfathers have either five or nine Tubular Chimes. In addition to quarter hour chimes, all Herschede clocks have an hour strike — a deep, resonant tone that counts the passing hours.

5

Setting up the tubular bell clock

Your clock has arrived in four separate boxes (A,B,C&D). Contents of each box are noted below.

BOX A
1. Clock case
2. Door key (in small envelope attached to security pad around clock)
3. Finial (attached to side rail inside of case)

BOX B
1. Clock movement
2. Pendulum bob
3. Lyre (Model No. 250 only)
4. Suspension spring
5. Winding key
6. Four wood screws (in small envelope atop seat board)

BOX C
1. Chime tubes
2. Pendulum rod

BOX D
1. Chime weight (marked CT on bottom), 20 lbs. for 5 tube models; 26 lbs. for 9 tube models
2. Time weight (marked TT on bottom), 11 lbs.
3. Strike weight (marked ST on bottom), 11 lbs.

6

Check contents of boxes carefully before discarding. Some people inadvertently throw away the winding key and suspension spring with packing materials.

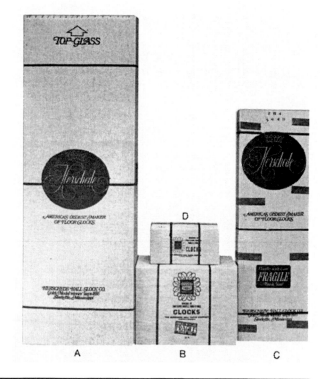

Tools for assembly

To assemble your clock, you will need the following tools and aids:
1. Wire cutters or pliers
2. Phillips screw driver
3. Spirit level
4. Pocket knife
5. Thin bladed screw driver
6. Shims (for leveling clock)
7. White cotton gloves or soft cloth (to use while handling brass parts and tubes)

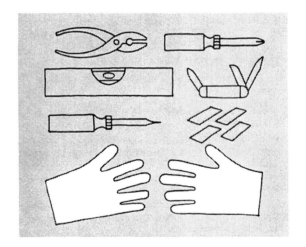

Unpacking

Unpack Box A first. Set upright and remove packing materials (don't forget door key). To get clock case out of box, grasp left side about midway up and lift from bottom. Once case is free, set it near its *permanent* location. Turn cabinet sideways and remove back panel by loosening four phillips screws. Remove top panel in same manner.

NOTE: All Herschede five and nine tube clock cabinets has removable tops, whether flat or curved. Remove retaining screws and top will lift off.

Next unpack the movement (Box B). Don't forget the small package containing the bob, winding key and suspension spring. Locate and lay aside for time being.

7

Installing movement

Remove protective tissue wrapping from movement, but DO NOT remove string holding chime hammers until movement is in case. Holding movement by the tube rack (Fig. 1), set seat board (A) on side rails (B) in cabinet. Slide movement forward until it is flush with face of cabinet. Line up clock face with opening in cabinet. Using key, open front door. Next, using a nail or long, thin object, line up screw holes in seat board and side rails. Screw seat board to side rails, using four phillips screws located in small envelope on top of seat board. Screws should be *tight*.

8

Fig. 1

Hanging chime tubes

Open Box C containing chime tubes. A white ticket identifies the strike (hour) tube in both the five and nine tubular bell movements. The pendulum rod is taped to the side of this tube. Gently remove rod (it is very delicate) and set aside. Unwrap the plugged (solid) end of the strike tube, exposing a small brown string. (Do not unwrap rest of tube at this time. This would only expose it to fingerprints.) Insert strike tube through front of cabinet and lean against cabinet wall. Remove protective string holding hammer springs. Cut and remove *white cords only*. Do not cut black strings — they are part of the chime system. Read red tag. It warns you not to tamper with adjustment screws on hammer springs (Fig. 2). They were pre-set at the factory. You are now ready to hang the strike (hour) tube. In the *five tube movement*, the hour tube should be hung on the extreme right button, as seen from back (Fig. 3). The rest of the tubes should be hung from *right to left* (as seen from back), starting with the longest chime tube and proceeding to the shortest. In the *nine tube movement*, the hour tube hangs on the second slot of the hour tube hanger. This hanger is located on the left side of the rack (Fig. 4). Getting there with the hour tube is a little tricky. First slide the lower end of the tube to the left back corner of the cabinet. Next slide the upper end under the chime rack on the extreme right (Fig. 5). Now slide the tube along the back of the seat board to the extreme left and hang on the hour tube hanger. Hang rest of the tubes on buttons at

Fig. 2

Fig. 3

Fig. 4

back of the tube rack (Fig. 3), starting with the longest tube on the right (as seen from back) and proceeding to the shortest. See page 15 also for layout. Replace back of cabinet and position clock in *permanent* location.

Move to front of cabinet and slide wrapping material from chime tubes. Pull gently from bottom of tube and "snake" off. (NOTE: Wear white cotton gloves or use a soft cloth when handling chrome or brass parts. Moisture in your hands will tarnish metal. Wipe off fingerprints immediately if you accidentally touch something you shouldn't.)

Fig. 5

9

Leveling cabinet

Place a spirit level alongside cabinet from front to back and side to side. See pictures below.

Problems can result from imperfect leveling. Check the balance periodically — especially if the clock is on carpet and may settle after original leveling.

Assembling pendulum

Assemble pendulum (Fig. 6). Remove adjusting nut (A) from rod (B). Slide rod through bob (C), making sure polished side of bob is on same side as the beat adjustment disc (D). (NOTE: On Model No. 250 insert lyre (Fig. 7) on pendulum rod before installing bob.) Replace adjusting nut. Screw on until top of bob reaches scratch mark on back of pendulum rod. Remove suspension spring screw (E). Insert suspension spring (F) in slot (G) and replace screw. Screw should be snug but not tight. Install pendulum in clock through front door. Hang T-bar in anchor bridge (Fig. 8). Insert beat adjusting pin in anchor fork (Fig. 9).

10

Front to Back   Side to Side

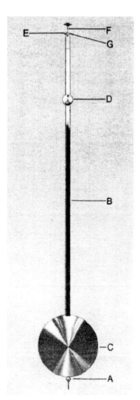

Fig. 6

Fig. 7

Fig. 8

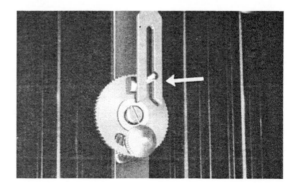

Fig. 9

11

Hanging weights

Now you are ready to hang the weights. Open Box D and remove wrapping from weights (leave brown paper until weights are hung.) Hang the large weight first. It goes on the right cable (facing front of clock). Stamped CT (for Chime Train) on bottom, this weight powers the chime gears. It weighs 20 pounds in a five tube movement and 26 pounds in a nine tube movement. The other two weights weigh 11 pounds each. Hang the one stamped TT (for Time Train) in the center (it powers the pendulum) and the one stamped ST (for Strike Train) on the left (it powers the hour strike). To install weights, flip pulley (Fig. 10) and hang on stirrup. Remove brown paper. NOTE: On clocks with an electric movement, hang weight pulley cords on hooks under seat board, then hang weights on pulley. (The weights are dummies and have no actual bearing on the operation of the clock.) Plug into a standard 110 volt electrical outlet.

Fig. 10

Correct chime selector positions

If chime selector is not correctly positioned, a malfunction or stoppage during playing of chime may occur. Do not turn chime selector lever when clock is chiming. Turn only after chimes have finished playing. Failure to do so may cause bent or broken pins on chime cylinder.

12

Westminster

Canterbury

Whittington

Fig. 13

Setting clock

Push strike silent lever up to strike position (Fig. 11). Turn chime indicator counterclockwise to chime (Fig. 12). Select chime (Fig. 13). Insert winding key in arbors (Fig. 14) and wind weights until top of weights are even with top of door opening. Check cable to see that it tracks properly in the winding drum grooves (Fig. 15). Swing pendulum gently to the left and let go. Listen to the beat. It should have an even, regulated tick and swing evenly in both directions. Rarely does a tubular bell clock arrive out of beat. If it needs adjusting, however, loosen the beat adjusting nut (Fig. 16, A) and move the beat adjusting disc (B) either left or right (using trial and error method) until the beat becomes regular. Tighten nut.

Fig. 11 Shown on Silent

Fig. 12 Shown on Silent

Fig. 14

Fig. 15

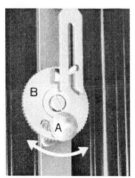
Fig. 16

13

Checking chimes

Turn the minute hand clockwise to the quarter-hour. Let chime. Turn hand to half-hour (Fig. 17). Let chime. Turn hand to three-quarter hour. Let chime. Turn hand to 12 o'clock. Let chime and strike hour.

Setting time

You are now ready to set the time. Do so by moving the minute hand clockwise (hour hand will follow). Be sure to stop at each quarter hour and let the clock go through its chime sequence. **Failure to do so could result in a broken pin inside the movement.** Caution: if clock runs down completely due to failure to wind, be sure to wind *before* setting hands. This allows chime to complete its cycle.

Setting moon dial

Position moon directly beneath numeral 15. Using an almanac as your reference, count the number of days past full moon. Turn moon dial (Fig. 18) two clicks clockwise (facing dial) for every full day (24 hours) past full moon. For example, if moon is five days past full moon you would move dial 10 clicks clockwise to set.

Replace cabinet top, screw in finial and your clock is set for many years of dependable service.

14

Fig. 17

Fig. 18

Position of tubular bells
From left to right, facing front of clock

5-tubes
1-1/8" overall diameter

HOUR TUBE   4   3   2   1

9-tubes
1-3/8" overall diameter

8   7   6   5   4   3   2   1   HOUR TUBE

9-tubes (250 model only)
1-1/2" overall diameter

8   7   6   5   4   3   2   1   HOUR TUBE

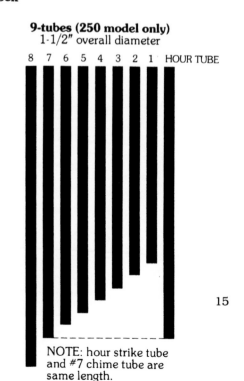

NOTE: hour strike tube and #7 chime tube are same length.

15

Regulating timekeeping

Changing the speed of timekeeping on any Herschede pendulum clock is accomplished by moving the pendulum bob up and down.

Faster: To speed up timekeeping, turn pendulum rating nut (Fig. 25) to right (clockwise) to raise pendulum bob.

Slower: To slow down timekeeping, turn pendulum rating nut (Fig. 25) to left (counter clockwise) to lower the pendulum bob.

This adjustment should be made gradually. Check timekeeping when you wind clock and make adjustment accordingly. Use the same clock or watch each time you check the clock, this will minimize error caused when using different sources of correct time.

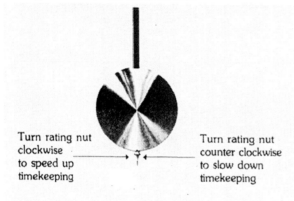

16

Turn rating nut clockwise to speed up timekeeping   Turn rating nut counter clockwise to slow down timekeeping

Care and maintenance

Your Herschede clock requires very little attention. There are a few things you can do, however, to increase its life considerably.

1. Wax and polish your clock cabinet as frequently as you wax and polish your other fine furniture. Any good furniture polish or paste wax will do.

2. Oil your clock every two or three years, using only good clock oil. Clean it every five to ten years depending on climate conditions. Extremely arid or salty air, heat or cold may necessiate more frequent servicing. Desert climate with high heat and low humidity causes oil to dry up. Salt air causes oil to break down. Note: Unless you are mechanically skilled, we don't recommend your attempting to clean your clock. Instead call a reputable clock service station (see list of recommended service stations in your owner's kit).

A few do's and don'ts

DO wind your clock every seven or eight days. If weights run down and pendulum stops, wind weights partly up. Set clock, then wind weights slowly to top and gently swing pendulum.

DO hang weight marked CT (underneath the weight) on the right (facing the clock). Otherwise, clock won't operate properly.

DO see that your clock is level at all times.

DO remove key from door of grandfather clocks to prevent tampering.

DO check contents of cartons thoroughly before discarding. Otherwise, you might inadvertently throw out suspension spring and winding key with packing materials.

DO have a trained clock serviceman oil and clean your clock at recommended intervals (see Care and maintenance). We recommend a good synthetic clock oil such as Nye or Mobis brands.

DO allow clock to go through its chime sequence when advancing minute hand. Failure to stop at each quarter hour could result in a broken pin inside movement.

DO specify clock model and type movement (five tube or nine tube if tubular bell) if you have occasion to write the factory about a problem.

DO stop pendulum swinging when leaving for more than a week.

DON'T touch pendulum bob or weights with your bare hands. Perspiration contains acid that causes weights, shells and pendulum bob to tarnish. Wipe immediately with soft, clean cloth if touched.

Use extreme caution when cleaning glass inside clock. Spray the cleaner on cloth and then wipe glass. Do not spray directly into cabinet. Also, do not spray move-

ment with any type of lubricant, as this will void your warranty.

DON'T locate clock near heating or air conditioning vents. Excessive humidity could affect cabinet or movement.

What to do if problems occur

Your Herschede clock should repay you with accurate timekeeping if it receives regular maintenance.

Should you ever encounter difficulty in regulating your clock or have trouble with the movement, be sure to contact an Authorized Herschede Service Station as soon as possible.

17

NOTES ON SETTING UP THE CLOCK

There is much more to the setup of a Herschede tubular bell clock than just getting all the pieces together. You will want to assemble everything smoothly, with a minimum of backtracking, tube banging, and doing things over. Another important aim is to identify any last minute adjustments you need to make. No matter how good your overhaul might have been, a sour chime note will mean a return service call at no charge. This section highlights the key areas for you to watch during the setup.

The procedure which follows is for a 9-tube clock of more recent vintage. Older clocks may have to be handled differently; interior and exterior cabinet dimensions vary considerably. This can affect the way you can remove and install the tubes.

Placing the Cabinet and Preliminary Leveling

When you are ready to begin the setup, stand the clock cabinet in the exact location it is to occupy, and have the customer approve this placement. Explain that it is difficult for you to move and level the clock again after all your work is done and the tubes are hanging in the cabinet. Allow the customer a few moments to decide on the location of the clock. This will make it less likely he or she will ask you to move and level the clock later.

Older clocks do not have adjustable leveling screws, so you must place shims under the corners to level the cabinet. Use squares of thin wood, not cardboard. All the newer clocks have adjustable feet. The two types are shown in Figure 77. Both consist of a threaded insert driven into a hole in the cabinet base (one at each corner). A screw is turned into the insert. One screw is the spike type (shown with the threaded insert attached), intended for thick carpets. The other type has a rounded plastic base. Whether the clock is to stand on a hard floor or a

Fig. 77. Clock feet: spiked type (left) to be avoided; and plastic-tipped style (right) preferred.

carpet, I recommend the plastic-tipped feet. I have found them to work well on any floor covering. Spikes are risky to use, because of the danger you will tear the carpet as you pivot the cabinet on one corner, to move it out from the wall.

The leveling feet will not have any effect unless they are turned out far enough to sink into the carpet. The cabinet must stand on these feet, not on the bottom cabinet molding. On a hard floor, the feet need only be far enough out so you can adjust them. Check the cabinet with your level; the clock must not lean forward or back, or to the side. Turn the leveling feet in or out to raise or lower them. On a carpeted installation, gently rock the cabinet from front to back to get the "feel" of it. If one foot is not down far enough, the cabinet will tip in that direction. Determine which foot to adjust, turn it, and re-check.

Installing the Movement

When you are satisfied with the leveling, tilt the clock on one front foot and pivot it out from the wall. The idea is to gain access to the rear of the clock. When you pivot the clock back later it may not need any re-leveling, but it should be checked. Install the movement, seatboard, tube rack, and dial from the rear, taking care not to catch the crutch (the fork) on the back of the cabinet as you go in. Locate the seatboard screw holes; insert and tighten all four wood screws. It is a good idea to check to be sure the clock door will close properly. You might have to make yourself a note to loosen the two screws which fasten the movement to the seatboard later, to permit a slight adjustment of the movement position. It is not an easy adjustment to make because the wooden pulley blocks are in your way.

Remove any packing materials from the movement now. Your access to these items will not be as good later on. Leave the wooden pulley blocks in place, of course. They don't come off until the weights are hung, near the end of the setup.

Hanging the Tubes

One approach is to put all nine tubes in the clock through the back panel, then move the clock into position. The risk is that you will bang and clash tubes together. This may not do any harm, but it upsets the customer. If you try to pack the tubes to prevent banging, it's still awkward.

Another procedure seems to work much better. To begin, install the hour tube from the back, leaning it against the tube rack in its approximate location. Then follow with the next two longest tubes, #8 and #7. These can also be leaned against the rack. Incidentally, the hour tube and tube #8 are hard to tell apart. You can measure one against the other; the hour tube is longer at 62 inches versus 61-5/8 inches for #8. On "The Clock", No. 250 only, the hour tube is exactly the same length as tube #7.

At this point, install the back panel of the cabinet with the four wood screws. The clock can now be pivoted back to its original position where it is to run. With the three tubes leaning in the case, you can still tilt the clock enough to adjust the clock feet again if the case is not quite level. Now hang the hour tube, tube #8, and tube #7, which are already in the case. Follow by working #6 and #5 in through the front of the clock. Remember, "longest on the left" is the rule for hanging the Herschede tubes. After you have hung #4, do #1, the shortest, on the far right. Then hang #3 and #2. It is easier if you don't have to hang the far right hand tube last, because of limited space. When all the tubes are in place on the tube rack, look at how they hang. Two may touch near the bottom, leaving a gap between the next two tubes. Correct this by lifting up on the tube and pulling the cord slightly. Tubes that touch each other will make strange noises in the night!

Starting the Clock

Hang the three weights, being especially careful with the heavy chime weight. Do not forget to remove the three wooden pulley blocks. They do a good job keeping the cables from getting tangled, but now they must come off. Don't leave them for the customer. Before you can remove the blocks, you must create some slack in the cables. Remove the dial, and start the fork ticking back and forth with the pendulum off. Turn the hands to make the clock chime and strike. After a short time, the strike and chime pulley blocks will be loose enough to remove. To take out the time block, you have to take off the weight, tilt the pulley until the cable comes off the wheel, and then remove the wooden block. You must keep some tension on the cable to prevent tangling. Put the cable back on the pulley, and hang the weight again.

Now install the pendulum and set the beat. Figure 78 shows the beat adjusting disc and nut on the pendulum shaft. If necessary, loosen the nut to enable you to move the disc. The nut must be tight, however, when you are done. The nut should be reasonably well centered in the slot in the disc. If you have to move the disc all the way to one end to set the beat, try reaching up to adjust the anchor. It can be moved if you hold the anchor as you move the crutch, but don't "bottom out" the anchor in the escape wheel. Then adjust the beat with the disc on the pendulum. Always set the beat with the smallest possible pendulum swing; overswing can disguise an out-of-beat condition.

Fig. 78. The beat setting mechanism consists of a beat adjusting disk with a crutch pin and a knurled nut. The pallet fork is shown on each side of the pin.

Final Adjustments

Listen to the chimes. If one or more notes sound too loud or too soft in relation to the other notes, there are several things you can do. First, check the clearance of the hammer head to the tube. Decrease or increase the clearance by turning the adjusting screw near the top of the chime lever, shown in Figure 79. You cannot reach the cord adjusters for the hammers corresponding to the middle tubes unless the dial is still off. Another adjustment is a thumbscrew on each of the hammer springs. A red tag was always attached to the tube rack on a new

Fig. 79. Chime lever cord adjuster, new style. The older style used a clamping arrangement that was harder to adjust.

clock, warning the customer not to tamper with these screws. The point is that if they are too tight, the clock will stall because of the increased tension. Adjust the screws if you feel they need it, and especially if you feel they may have been touched before.

This chapter concludes with a Herschede instruction page from the 1980's. The instructions cover the procedure for removing a 9-tube Herschede movement from the clock cabinet and packing it in an original factory shipping box (the box is item B on page 64). Although the boxes are no longer available, the points on this page will be helpful when a movement is packed for shipment.

The address shown on the page is not the current Herschede address.

Herschede Hall Clock Company
Hwy. 12 West, P.O. Box 825
Starkville, MS 39759

PACKING INSTRUCTIONS FOR 9 TUBE MOVEMENT

1. Turn chime selector indicator hand to Westminster position.

2. Move strike and chlme levers to silent positions.

3. Place cable blocks on pulleys, wind cable up snug, holding block straight so top notch fits over movement mounting channel.

4. Remove weights and pendulum.

5. Remove back cloth covered panel (4 corner screws).

6. Remove chime tubes.

7. Remove 4 seatboard screws.

8. Slide movement with seatboard out back, lifting slightly to clear anchor fork.

9. Interlace hammers with string from one side of chime rack to other. (This step is very important to prevent damage.)

10. Tie pendulum fork to prevent excessive vibrating in transit.

11. Put tissue paper behind hands to prevent scratching dial.

12. Wrap movement in brown paper.

13. Place movement in carton with chime rack down.

14. Put sheet of foam rubber on top of movement.

15. Close carton, tie, tape and or band.

16. Attach return label to top of carton.

Fig. 80. A Herschede procedure for removing and packing a 9-tube movement.

14

MOVEMENT DRAWINGS AND PARTS LIST

This chapter is a reference guide for the most common 9-tube "standard" Herschede movement. The design changed very little over many years. Parts are not always interchangeable, but many parts are similar on movements made decades apart.

Use the drawings, Figures 81-83, to help you locate and identify the major parts and assemblies. The drawings can also be used as a guide in assembling most 5 or 9-tube Herschede movements, whether new or old. Herschede part names and numbers are used throughout. Only major parts and assemblies are included in these three drawings. Assemblies contain pins, levers, screws, and other components. For example, the strike train main shaft assembly, part 2610, is made up of many separate parts. In addition to the wheel and drum, there is a click screw, washer, and other items.

Figures 84-91 are illustrated parts lists for the 9-tube movement. The information is reproduced by special arrangement with Herschede Hall Clock Company, the present-day supplier of parts for the tubular bell movements. Prices have been deleted from these lists, as they are subject to change. Please note that in 1980 new tooling was purchased by Herschede, and that parts made from the new tooling have a 5-digit number. For example, old #2630 becomes new #22630. In addition, the prefix HER denotes current Herschede item numbers for ordering purposes. To purchase parts, repairers should contact Herschede at the following address:

Herschede Hall Clock Company
3313 Harlan Carroll Rd.
Waynesville, OH 45068
phone (513) 897-5015

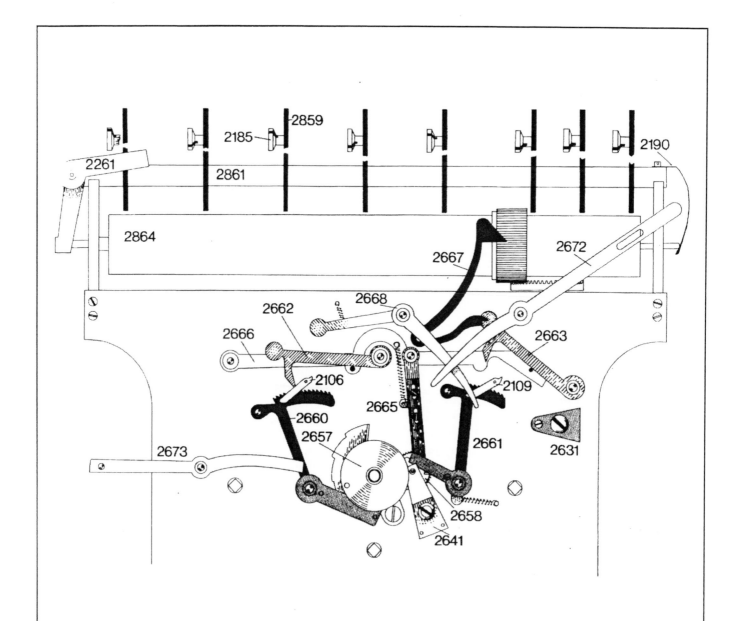

Fig. 81. Front view, 9-tube movement

2106 strike gathering pallet
2109 chime gathering pallet *
2185 chime lever cord adjuster
2190 hammer bar spring
2261 shift lever
2631 second shaft bridge
2641 intermediate bridge
2657 hour tube assembly
2658 intermediate shaft assembly
2660 hour rack
2661 quarter rack

2662 hour rack hook
2663 quarter rack hook
2665 lifting lever assembly
2666 long lever
2667 self adjusting lever assembly
2668 double lever assembly
2672 chime silent lever
2673 hour silent lever
2859 chime lever assembly
2861 9T hammer bar assembly
2864 cylinder 9T

*The chime gathering pallet 2109 is now supplied
in the offset version 22109. See pages 58 and 82.*

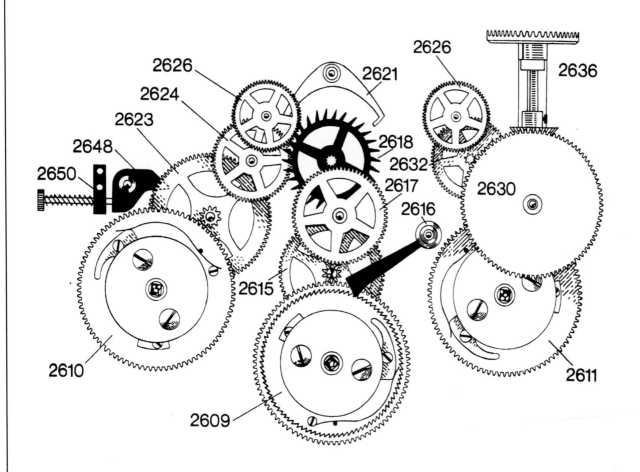

Fig. 82. Internal view, 9-tube movement

2609 T.T. main shaft assembly
2610 S.T. main shaft assembly
2611 C.T. main shaft assembly
2615 center shaft assembly (center wheel)
2616 maintaining shaft assembly
2617 3rd shaft assembly
2618 escape shaft assembly
2621 anchor shaft assembly

2623 pin wheel shaft assembly
2624 S.T. gathering shaft assembly
2626 warning shaft assembly
2630 C.T. 2nd shaft assembly
2632 C.T. gathering shaft assembly
2636 crown wheel shaft assembly
2648 hour strike shaft assembly
2650 hour strike bridge assembly

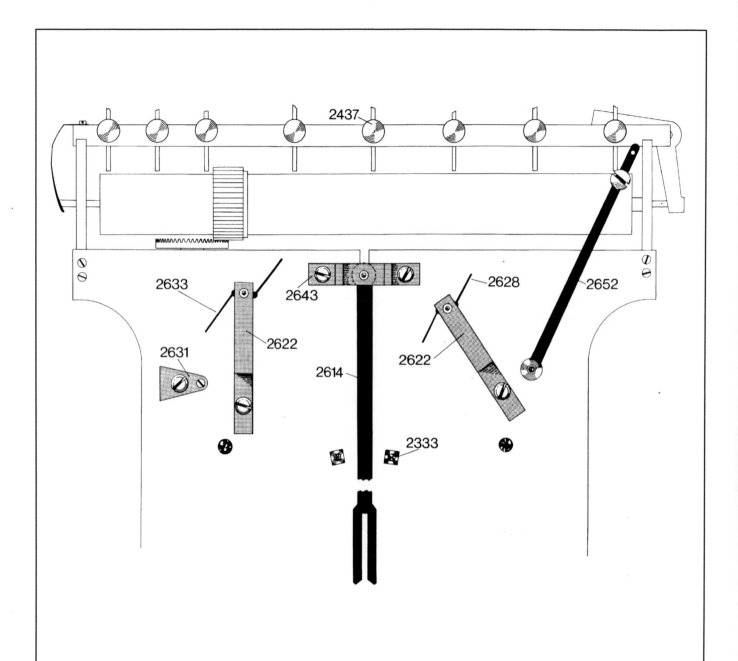

Fig. 83. Rear view, 9-tube movement

2333 fork banking pin
2437 chime lever adjust screw
2614 fork
2622 fan bridge
2628 S.T. fan shaft assembly
2631 second shaft bridge
2633 C.T. fan shaft assembly
2643 anchor bridge
2652 hammer arm

HERSCHEDE WEIGHTS

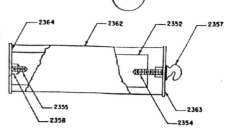

FOR TUBULAR BELL CLOCKS

9 TUBE WEIGHT ASSEMBLY

HER-2735
Complete set
(Includes: 1 #2732 chime train weight & 2 #2734 (1 strike & 1 time train) weights.
HER-2735WO/WT
Set of 3 weights
(No insert)
HER-2736
Set of 3 weights
for 5 tube clock.

CHIME TRAIN WEIGHT

Part#	Description
HER-2732	Complete weight
HER-2732WS	Weight Shell
	No insert

Part#	Description	Measurement
HER-2350	Lead Weight Insert	26 lbs
HER-2356	Weight Shell Rod	.250" Dia. 8.625" Long
HER-2357	Shell hook	
HER-2358	Shell nut	
HER-2359	Large Weight shell	3.502" Dia. 8.750" Long
HER-2360	Large Upper lid	3.631" Dia.
HER-2361	Large Lower lid	3.631" Dia.

STRIKE & TIME TRAIN WEIGHTS

Part #	Description
HER-2734	Complete weight
HER-2734WS	Weight shell
	No insert.

Part#	Description	Measurement
HER-2352	Steel Weight Insert	11 lbs
HER-2354	Weight Stud Long	1 1/4" L
HER-2355	Weight Stud Short	7/8" L
HER-2357	Shell hook	
HER-2358	Shell nut	
HER-2362	Small Weight Shell	2.501" Dia. 8.75 L.
HER-2363	Small Upper Lid	2.641" Dia.
HER-2364	Small Lower Lid	2.641" Dia.

HERSCHEDE TUBES

TUBE SET OF 5, NICKEL PLATED 1 1/8" OUTSIDE DIAMETER

ITEM	LENGTH	PART #	
Tube 1	42 15/16"	HER-2413	
Tube 2	45 1/2"	HER-2416	
Tube 3	48 5/8"	HER-2418	
Tube 4	56 1/2"	HER-2419	
Hour tube	58 3/16"	HER-2411	1 3/8" diameter
SET of 5		HER-2421	

TUBE SET OF 9, NICKEL PLATED 1 3/8" OUTSIDE DIAMETER

ITEM	LENGTH	PART #
Tube 1	43 1/64"	HER-2401
Tube 2	44 1/2"	HER-2402
Tube 3	47 1/4"	HER-2403
Tube 4	49 15/16"	HER-2405
Tube 5	53 3/8"	HER-2406
Tube 6	55 3/16"	HER-2407
Tube 7	58 3/16"	HER-2411
Tube 8	61 5/8"	HER-2412
Hour tube	62"	HER-2408
SET OF 9		HER-2420

The above tubes are of the new type which is suspended from one post.
If the old style is needed, please call for availability & price.
If other tubes are needed, please call for availability & price.

NEW TYPE   OLD STYLE

Fig. 84

 **HERSCHEDE 3 WEIGHT TUBULAR BELL MOVEMENT PARTS**

Her-2006 Pivot Eccentric Bushing

HER-2019 Upper Movement Post *Req 2

HER-2020 Lower Movement Post *Req 2

HER-2021 Spreader Post

HER-2022 Hour Bridge Screw

HER-2024 Post Screw

HER-2025 Set Screw Long

HER-2026 Set Screw

HER-2027 Dial Post Knob Screw & Hammer Bar Spring Screw

HER-2044 Click Screw T.T.

HER-2045 Maintaining Wheel

HER-2047 Maintaining Spring

HER-2050 Click Screw C.T. & S.T.

HER-2054 Click Spring Fastening Screw *Req 3

HER-2062 Main Wheel Spring Washer *Req 3

HER-2063 Main Washer Screw *Req 3

HER-2064 Washer (Teflon)

HER-2086 Anchor

HER-2087 Anchor Bushing

HER-2107 Cylinder Gear Screw Assembly

HER-2117 Fan Wing Screw *Req 2

HER-2130 Second Shaft Adjusting

HER-2154 Crown Wheel Bridge

HER-2155 Crown Bridge Adjusting

HER-2161 Spreader Post Screw

HER-2165 Cylinder Pin (Steel) *Req (9T 90 5T 20)

HER-2166 Cylinder Self adjusting Pin (Steel)

HER-2171 Cylinder Pin (Brass)

HER-2189 Hammer Bar Fastening Screw

HER-2201 Cylinder Bridge Screw 5T *Req 4

Pictures not actual Size

Fig. 85

HERSCHEDE
3 WEIGHT TUBULAR BELL
MOVEMENT PARTS

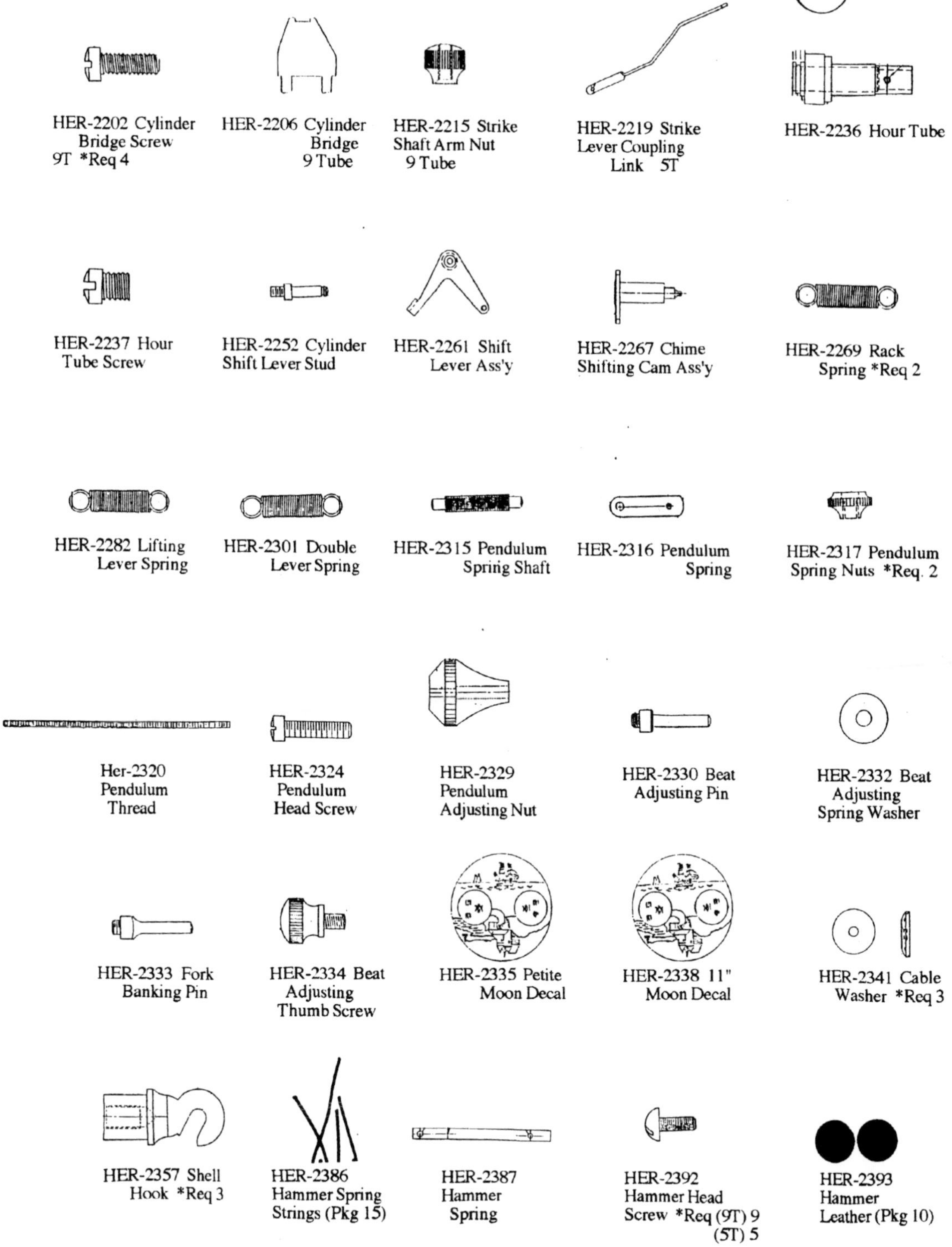

HER-2202 Cylinder
Bridge Screw
9T *Req 4

HER-2206 Cylinder
Bridge
9 Tube

HER-2215 Strike
Shaft Arm Nut
9 Tube

HER-2219 Strike
Lever Coupling
Link 5T

HER-2236 Hour Tube

HER-2237 Hour
Tube Screw

HER-2252 Cylinder
Shift Lever Stud

HER-2261 Shift
Lever Ass'y

HER-2267 Chime
Shifting Cam Ass'y

HER-2269 Rack
Spring *Req 2

HER-2282 Lifting
Lever Spring

HER-2301 Double
Lever Spring

HER-2315 Pendulum
Spring Shaft

HER-2316 Pendulum
Spring

HER-2317 Pendulum
Spring Nuts *Req. 2

Her-2320
Pendulum
Thread

HER-2324
Pendulum
Head Screw

HER-2329
Pendulum
Adjusting Nut

HER-2330 Beat
Adjusting Pin

HER-2332 Beat
Adjusting
Spring Washer

HER-2333 Fork
Banking Pin

HER-2334 Beat
Adjusting
Thumb Screw

HER-2335 Petite
Moon Decal

HER-2338 11"
Moon Decal

HER-2341 Cable
Washer *Req 3

HER-2357 Shell
Hook *Req 3

HER-2386
Hammer Spring
Strings (Pkg 15)

HER-2387
Hammer
Spring

HER-2392
Hammer Head
Screw *Req (9T) 9
(5T) 5

HER-2393
Hammer
Leather (Pkg 10)

Pictures not actual Size

Fig. 86

HERSCHEDE
3 WEIGHT TUBULAR BELL
MOVEMENT PARTS

HER-2398
Hammer Spring
Adj. Screw
*Reg (9T) 9
(5T) 5

HER-2399
Adj. Screw
Spring
*Reg (9T) 9
(5T) 5

HER-2429 Dial
Post Bushing
Screw

HER-2437 Chime
Lever Adj. Screw

HER-2440 Moon
Lever Screw

HER-2443 Moon
Shift Lever Ass'y

HER-2444 Flat
Head Dial Screw
Pkg. of 3

HER-2449 Moon
Stud

HER-2450 Moon
Stud Washer

HER-2454 Moon
Spring Screw

HER-2459 Moon
Bushing Screw

HER-2460 Dial
Screw-Flat Head
(Gold Plated)

HER-2461 Dial
Screw-Round
Head (Silver Plated)

HER-2468 Crown
Screw
Pkg. of 3

HER-2490 Indicator
Washer *Req 2

HER-2491
Indicator Nut
*Req 2

HER-2500 Hour
Hand 11"

HER-2503 Minute
Hand 11"

HER-2506 Second
Hand

HER-2507 Second
Hand Bushing

HER-2508
Second Hand
Screw

HER-2509 Hand
Washer

HER-2510 Minute
Hand Petite

HER-2513 Hour
Hand Petite

HER-2525 Movement
Fastening Screw

HER-2528 Petite
Corner Seroll
Screw

HER-2605 Dial
Post Assembly
*Req 4

HER-2614
Fork

HER-2622 Fan
Bridge Ass'y

HER-2627 Beat Adj.
Disc Ass'y

Pictures not actual Size

Fig. 87

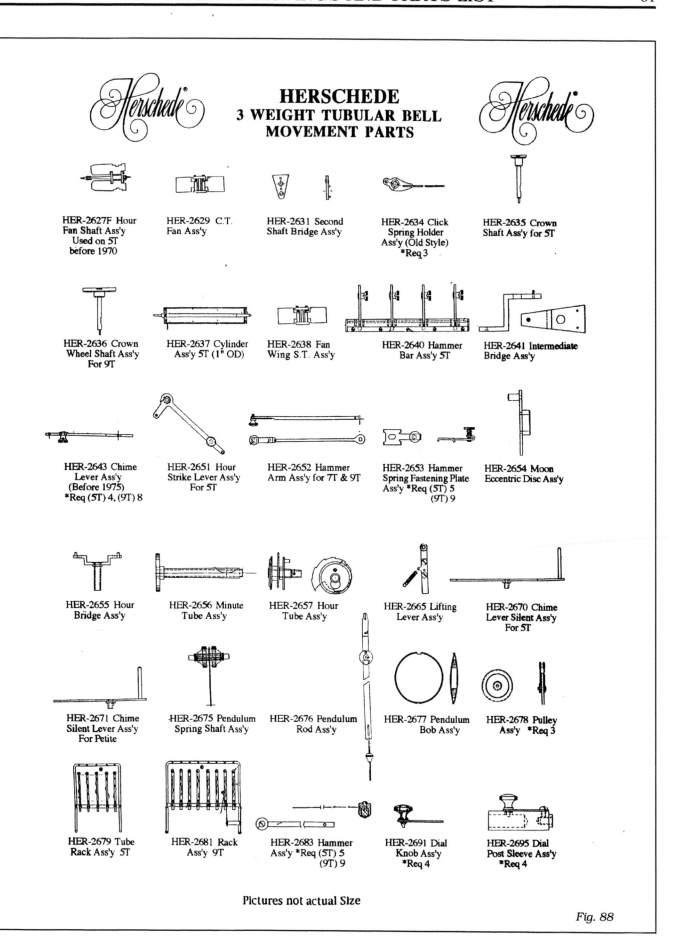

HERSCHEDE
3 WEIGHT TUBULAR BELL MOVEMENT PARTS

HER-2627F Hour Fan Shaft Ass'y Used on 5T before 1970

HER-2629 C.T. Fan Ass'y

HER-2631 Second Shaft Bridge Ass'y

HER-2634 Click Spring Holder Ass'y (Old Style) *Req 3

HER-2635 Crown Shaft Ass'y for 5T

HER-2636 Crown Wheel Shaft Ass'y For 9T

HER-2637 Cylinder Ass'y 5T (1" OD)

HER-2638 Fan Wing S.T. Ass'y

HER-2640 Hammer Bar Ass'y 5T

HER-2641 Intermediate Bridge Ass'y

HER-2643 Chime Lever Ass'y (Before 1975) *Req (5T) 4, (9T) 8

HER-2651 Hour Strike Lever Ass'y For 5T

HER-2652 Hammer Arm Ass'y for 7T & 9T

HER-2653 Hammer Spring Fastening Plate Ass'y *Req (5T) 5 (9T) 9

HER-2654 Moon Eccentric Disc Ass'y

HER-2655 Hour Bridge Ass'y

HER-2656 Minute Tube Ass'y

HER-2657 Hour Tube Ass'y

HER-2665 Lifting Lever Ass'y

HER-2670 Chime Lever Silent Ass'y For 5T

HER-2671 Chime Silent Lever Ass'y For Petite

HER-2675 Pendulum Spring Shaft Ass'y

HER-2676 Pendulum Rod Ass'y

HER-2677 Pendulum Bob Ass'y

HER-2678 Pulley Ass'y *Req 3

HER-2679 Tube Rack Ass'y 5T

HER-2681 Rack Ass'y 9T

HER-2683 Hammer Ass'y *Req (5T) 5 (9T) 9

HER-2691 Dial Knob Ass'y *Req 4

HER-2695 Dial Post Sleeve Ass'y *Req 4

Pictures not actual Size

Fig. 88

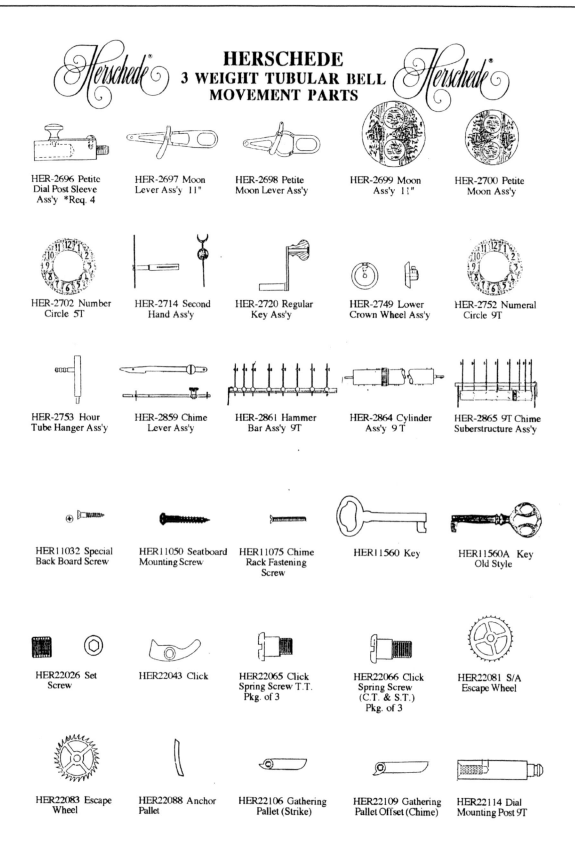

HERSCHEDE
3 WEIGHT TUBULAR BELL
MOVEMENT PARTS

HER-2696 Petite Dial Post Sleeve Ass'y *Req. 4

HER-2697 Moon Lever Ass'y 11"

HER-2698 Petite Moon Lever Ass'y

HER-2699 Moon Ass'y 11"

HER-2700 Petite Moon Ass'y

HER-2702 Number Circle 5T

HER-2714 Second Hand Ass'y

HER-2720 Regular Key Ass'y

HER-2749 Lower Crown Wheel Ass'y

HER-2752 Numeral Circle 9T

HER-2753 Hour Tube Hanger Ass'y

HER-2859 Chime Lever Ass'y

HER-2861 Hammer Bar Ass'y 9T

HER-2864 Cylinder Ass'y 9 T

HER-2865 9T Chime Suberstructure Ass'y

HER11032 Special Back Board Screw

HER11050 Seatboard Mounting Screw

HER11075 Chime Rack Fastening Screw

HER11560 Key

HER11560A Key Old Style

HER22026 Set Screw

HER22043 Click

HER22065 Click Spring Screw T.T. Pkg. of 3

HER22066 Click Spring Screw (C.T. & S.T.) Pkg. of 3

HER22081 S/A Escape Wheel

HER22083 Escape Wheel

HER22088 Anchor Pallet

HER22106 Gathering Pallet (Strike)

HER22109 Gathering Pallet Offset (Chime)

HER22114 Dial Mounting Post 9T

Pictures not actual Size

Fig. 89

 **HERSCHEDE**
3 WEIGHT TUBULAR BELL
MOVEMENT PARTS

HER22115 Flat
Head Screw for
Dial Ass'y
Pkg of 8

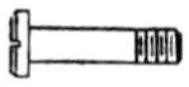

HER22120 Anchor
Bridge Screw (SA)
Pkg. of 3

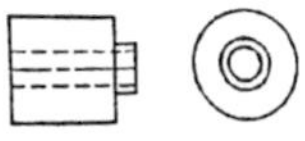

HER22121 Anchor
Bridge Bushing (SA)

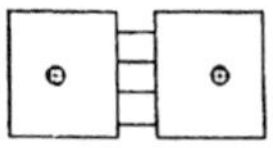

HER22128
Suspension
Spring (SA)

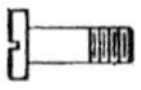

HER22226 Hour
Bridge Screw

HER22270 Rack
Arm Spring Screw

HER22488
Indicator Hand

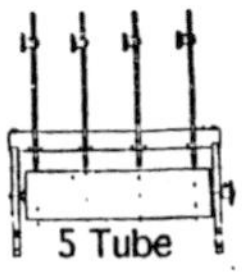

5 Tube

HER22556
Superstructure
Ass'y 5T

HER22609 Main
Shaft Ass'y T.T.

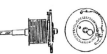

HER22610 Main
Shaft Ass'y S.T.

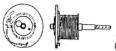

HER22611 Main
Shaft Ass'y C.T.

HER22612 Anchor
Ass'y

HER22615 Center
Shaft Ass'y

HER22616
Maintaining Shaft
Ass'y

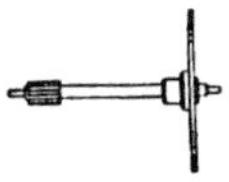

HER22617 Third
Shaft Ass'y

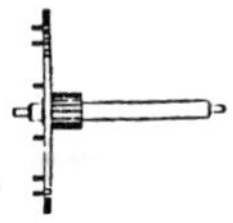

HER22623 Pin
Wheel Shaft Ass'y

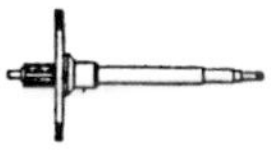

HER22624 S.T.
Gathering Shaft
Ass'y

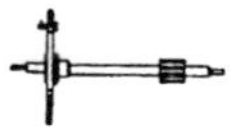

HER22626 Warning
Shaft Ass'y

HER22628 S.T.
Fan Shaft Ass'y

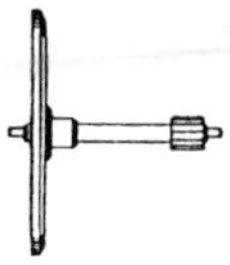

HER22630 C.T.
Second Shaft Ass'y

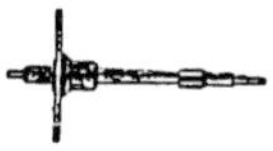

HER22632 C.T.
Gathering Shaft
Ass'y

HER22633 C.T.
Fan Shaft Ass'y

HER22638 5 Tube
Cylinder Ass'y
1 1/8" OD

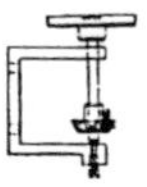

HER22641 9 Tube
Crown Wheel
Bridge Ass'y

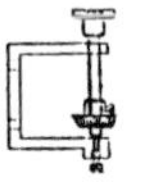

HER22642 5 Tube
Crown Wheel
Bridge Ass'y

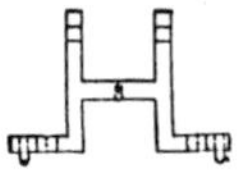

HER22643 Anchor
Bridge Ass'y S/A

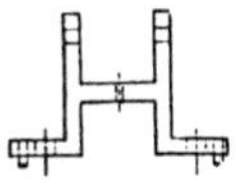

HER22644 Anchor
Bridge Ass'y

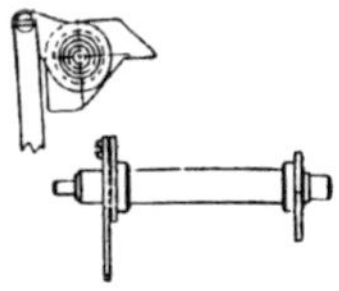

HER22645 5 Tube
Hour Strike Shaft
Ass'y

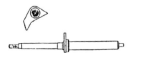

HER22648 Hour
Strike Shaft Ass'y

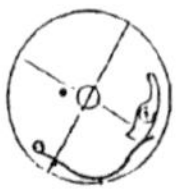

HER22649 Click
Ass'y S.T.

Fig. 90

HERSCHEDE
3 WEIGHT TUBULAR BELL MOVEMENT PARTS

HER22650 Hour Strike Bridge Ass'y   HER22652 Hammer Arm Ass'y   HER22655 Hour Bridge Ass'y   HER22658 Intermediate Shaft Ass'y   HER22659 Click Ass'y T.T.

HER22660 Hour Rack Ass'y   HER22661 Quarter Rack Ass'y   HER22662 Hour Rack Hook Ass'y   HER22663 Quarter Rack Hook Ass'y   HER22664 Click Ass'y C.T.

HER22666 Long Lever Ass'y   HER22667 Self Adjusting Lever Ass'y   HER22668 Double Lever Ass'y   HER22672 Chime Silent Lever Ass'y 7T & 9T   HER22673 Hour Silent Lever Ass'y 7T & 9T

HER22692 Quarter Rack Hook Ass'y 5T   HER22705 Silent Shifting Cam Ass'y   HER22722 Click Spring

**Pictures not actual size.
**Ass'y = Assembly
**OD = Outside diameter
**SA = Self Adjusting
**C.T. = Chime Train
**S.T. = Strike Train
**T.T. = Time Train

9T DIAL ASSEMBLY
HAND ENGRAVED - HER-2692

Used on all 9-Tube Clocks except Models #250 & 276
SIZES & STYLES OF DIALS VARIED, PLEASE CALL FOR AVAILABLITY

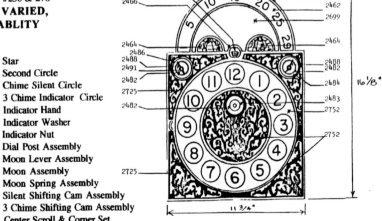

2031	Twist Pin bushing screw	2482	Star
2425	9 Tube Dial Back	2483	Second Circle
2429	Dial Post Bushing Screw	2484	Chime Silent Circle
2440	Moon Lever Screw	2486	3 Chime Indicator Circle
2449	Moon Stud	2488	Indicator Hand
2450	Moon Stud Washer	2490	Indicator Washer
2454	Moon Spring Screw	2491	Indicator Nut
2459	Moon Bushing Screw	2605	Dial Post Assembly
2460	Dial Screws - Flat Head	2697	Moon Lever Assembly
2461	Dial Screws - Round Head	2699	Moon Assembly
2462	Moon Circle	2701	Moon Spring Assembly
2464	Hemisphere (Set of 2)	2705	Silent Shifting Cam Assembly
2466	Crown Field	2707	3 Chime Shifting Cam Assembly
2467	Crown	2725	Center Scroll & Corner Set
2468	Crown Screw		(Hand Engraved - Set of 5 pcs.)
		2752	9-tube Numeral Circle Assembly

Fig. 91

15

REFINISHING THE PLATES AND WHEELS

Wouldn't it be ideal if a clock case could age gracefully, without being damaged and abused along the way? It does happen that way, sometimes, with the case giving the impression that it has been cared for, even if some cracks and fading have appeared here and there. All too often, however, a finish is ruined and refinishing is the only approach.

It's the same story with a clock movement such as a Herschede tubular bell. If it has been cleaned improperly and the finish completely ruined, polishing and lacquering are needed to restore the finish on the brass parts. This chapter describes a method for solving the special problems in refinishing the plates and the main wheels.

The featured clock is a Model 294 "Haverford" which I was told was only eight years old at the time it came in for repair. Someone had just done a disastrous cleaning job on the movement. Despite

Fig. 93. The restored Herschede 9-tube movement

its young age, it looked like it had come through both World Wars. Figure 92 shows a section of the back plate. There were scratches, stains, and pitted areas on both plates, and the polished pattern was completely defaced.

POLISHING THE PLATES

The circular ornamentation, called damascening, consists of interlocking rows of polished circles. It is a hallmark of Herschede tubular bell movements. Howard Klein, the owner of Herschede at the time, provided the basic concept for polishing the plates. He said that a hard felt lap, charged with grinding compound, could be used in a drill press to produce the circular pattern on the plates. Some kind of indexing grid system would be required to accurately locate the plate under the lap as the rows of circles were polished in the brass.

With this concept in mind, I worked out a method with the help of Ron Pfleger, a toolmaker. Ron punched out a set of leather laps, made the lap

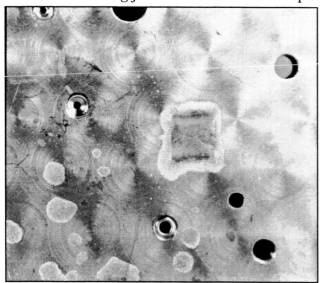

Fig. 92. A section of the damaged Herschede back plate

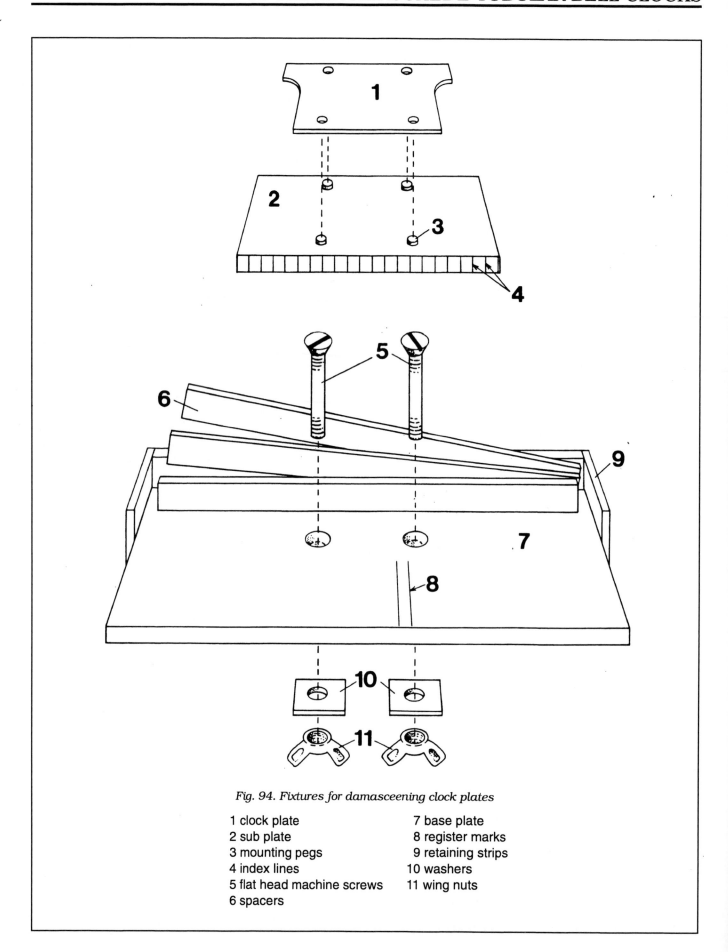

Fig. 94. Fixtures for damasceening clock plates

1 clock plate   7 base plate
2 sub plate   8 register marks
3 mounting pegs   9 retaining strips
4 index lines   10 washers
5 flat head machine screws   11 wing nuts
6 spacers

holder for the drill press, and came up with a way to index (space) the rows of polished circles. I built this indexing system for the drill press, then carried out the damasceening with very satisfying results.

Fig. 95. The base plate installed on the drill press

List of Materials and Equipment
- 1 pc. particle board 2' x 4' x 5/8" thick, or other flat, rigid board as available
- 3 pcs. pine molding 8' x 1" wide, selected to be as close as possible to 5/16" thick
- 1 pc. brass rod 1/4" dia. x 5" long
- 2 ea. flat head machine screw, 1/4-20 x 2"
- 2 ea. wing nuts for above
- 2 pcs. 3/16" aluminum or brass sheet 2-3" square
- 1 pc. steel rod 11/16" dia. x 3" long
- 8 pcs. leather laps 5/8" dia., approx. 1/8" thick (see text for more information)
- silicon carbide or other grinding compound
- contact cement
- drill press
- drills, sizes "B", 1/4", 17/64"
- countersink
- misc. saws, files, hand tools, etc.

Refer to Figure 94, facing page, to identify the parts. The sub plate (2) is a piece of the particle board cut to 9" x 7". Place the back plate of the clock (1) face up on this piece, with the bottom edge of the plate carefully aligned with one of the long edges of the sub plate. Clamp or fasten these two items with cloth tape in this position. Install a size "B" drill (.238") in the drill press chuck. Carefully use the four clock pillar holes as a template as you touch the spinning drill to the particle board. This marks the locations for the four mounting pegs (3). Remove the clock plate from the sub plate. Change

the size "B" drill for a 1/4" drill, and finish drilling the holes about .400" deep in the sub plate, to receive the pegs. Cut the pegs to approximately 1/2" long, then insert them into the holes. Place the clock plate over the pegs to lie flat on the sub plate. Check to be sure the pegs do not extend above the surface of the clock plate where they will damage the lap during the polishing operation. Correct the pegs by shortening them or drilling the holes deeper in the sub plate. Finish up the sub plate by marking the series of index lines (4) on the edge of the sub plate as shown, at 1/2" intervals.

For the base plate (7), cut a piece of particle board 12" x 24". Nail strips of molding across the 12" back and partially along the sides of the particle board as shown in Figure 94, to form the retaining strips (9). Next, drill and countersink two 17/64" clearance holes in the base plate for the flat head machine screws, spacing the holes to match the slots in your drill press table. Cut the pieces of 3/16" aluminum or brass sheet to form square washers about 1-1/4" square (or to match the underside of the drill press table). Drill a 17/64" clearance hole in the center of each washer for the flat head machine screw. Now place the base plate on top of the drill press table. Insert the screws into the holes in the particle board and through the slots in the table. Then install the square washers and the wing nuts on the screws. To finish up the base plate, draw a

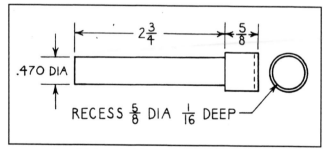

Fig. 96. Lap holder. The material is steel, and the dimensions are in inches.

pair of register marks (8) in the approximate location shown in Fig. 94. The marks are 9/32" apart and must be parallel to each other and to the sides of the base plate.

Cut nine spacers (part #6), each approximately 23-7/8" long, from the pine molding. These will be placed on the base plate to provide the spacing of the rows of polished circles. One of the spacers should be cut into three approximately equal pieces, for reasons that will be explained. In addition, one more spacer, of a thickness to be determined later, will be required. It may be made of a thin material such as cardboard.

The lap holder is made of 11/16" steel rod as shown in Figure 96. You can substitute brass, alu-

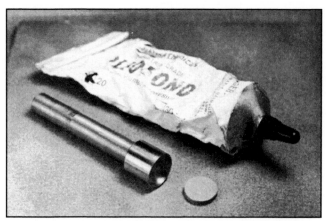

Fig. 97. Lap holder, leather lap, and contact cement

minum, or even wood for the holder. This particular design is based on the requirement that the holder fits in a drill press with a 1/2" capacity chuck. The recess is 5/8" diameter to receive the leather lap, which leaves only a 1/32" wall around the recess. The thin wall allows the lap to get as close as possible to obstructions on the front plate, which will be shown later.

The laps, stamped from leather approximately 1/8" thick, are a crucial part of the project. A sharp, clean piece of leather is required for good results in polishing. Several laps will be consumed in the process of polishing the plates. As illustrated in Figure 97, a lap is glued in place with contact cement. After the lap becomes worn or is damaged from contact with a sharp edge, pry it out of the recess in the lap holder. Clean out all the glue residue from the recess, wipe with alcohol, and glue in a new lap. This can be done in the middle of a polishing operation if necessary.

Silicon carbide grinding compounds are shown in Figure 98. I tried fine, medium, and coarse grades on a piece of scrap brass and did not see great differences in the results. The polished circles were slightly brighter with the 120 grit coarse compound, so I used that. Howard Klein suggested valve grinding compound, available in auto stores, for dama-

Fig. 98. Grinding compounds and leather laps

scening the plates. Also shown in Figure 98 are three new laps and a worn lap which had to be removed from the holder.

Preparing the Clock Plates

Disassemble and clean the movement, then finish all bushing work. To prepare the plates, remove all threaded parts. This includes the pillars and dial feet. Also tap out the long pin on the inside of the back plate. If you haven't already done so, remove the two brass screws and the crown wheel shaft assembly. When this is complete, the back plate will be clean. The front plate will retain a number of posts which are riveted in place. It is not practical to remove these.

The plates are now ready for removal of all scratches, pits, and original damascening. This is a time-consuming process. Most of the lacquer, if there

Fig. 99. A Herschede back plate is shown after it was polished to a matte finish. The plate is ready for damasceening.

is any left, can be removed in old clock cleaning solution. The rest will come off as the plate is treated with abrasives. Good results can be obtained with a piece of Scotch-Brite material, which produces a smooth, matte finish to mask small defects. I also experimented with wet/dry paper and kerosene. It is best to try different abrasives to see which one is most effective on the defects in your plates. Avoid abrasive grits coarser than 280, as these may add their own deep scratches which have to be removed by successively finer papers.

As I tried the damascening on various brass surfaces, I found that obvious scratches would show through, so these must always be removed. A smooth finish such as that shown on the plate in Figure 99 is a good foundation for the finished product. Wrap the abrasive paper around a piece of wood to present a flat surface to the brass plate. Take particular care not to round the edges of the plate. This is bad practice, anyway, but is especially harmful in this project; the lap may be unable to adapt to a curved edge. Polish the oil sinks in the plates. Polish as closely as possible around the posts which are riveted to the front plate.

Damascening the Plates

Install the lap holder, with lap glued in place, in the drill press chuck. Place all nine spacers (part #6) on the base plate against the rear retaining strip. The spacer which was cut in several pieces is placed closest to you. Wipe the outer surface of the clock plate with alcohol and then place it onto the mounting pegs (3) on the sub plate (2). Set the sub plate

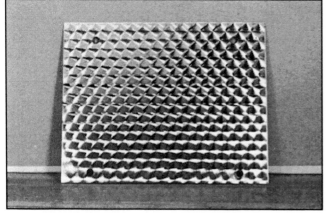

Fig. 100. Damascene a test piece first, for practice. Imagine removing 150 polished circles from the clock plate and starting over!

onto the base plate. The actual starting point for polishing the first circle is the strike-side lower corner for the back plate. The starting point for the front plate is the chime-side lower corner. Now loosen the wing nuts holding the base plate to the drill press table. Carefully slide the base plate front or back so that the center of the lap is near the bottom edge of the clock plate. The idea is to set the first row with the centers just on the plate. This will provide a row of partial circles for row #1. When the

Fig. 101. Working around a post on the front plate

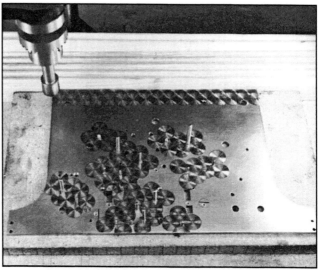

Fig. 102. The front plate after three rows have been damasceened. Note the pattern of circles around each post on the plate.

desired position for the base plate has been located, tighten the wing nuts firmly. Do not loosen the nuts again until both plates have been damascened, or the indexing of circles will be lost. The base plate would have to be lined up all over again.

Before we begin to polish the rows of circles into the plates, we must consider a special problem with the front plate. There are 15 or so posts that are riveted in place. It is not practical to remove them and then make new ones to install after the damascening is finished. I never seriously considered doing that, although I did think about eliminating the damascening entirely from the front plate. Ron Pfleger provided the solution to the problem. Polish a pattern of circles closely around each obstacle as shown in Figure 101. Then begin the damascening, row by row, from the bottom of the plate. Figure 102 shows the front plate after the third row has been damascened. When you reach a point where the lap cannot be brought in contact with the clock plate because of an obstacle, just skip that position and index to the next. In some areas of the plate, as many as four circles in sequence may be missed. However, the pattern around each obstacle fills in the damascened effect convincingly. Figure 93 shows the end result on the front plate.

Now back to the indexing procedure. The first circle of the first row is to be located with its center near the corner of the plate as described earlier. Do this by lining up an index line on the sub plate with one of the two register marks on the base plate. Locate the circle so that almost three quarters of it will be off the plate. You may find the side edges of the clock plate will tear up the leather of the lap. Ron suggested bringing a piece of brass of the same thickness as the plate up against the edge to support the lap and reduce wear. It is important to note that the register mark you have selected remains the mark used for row #1. Spread a small amount of compound on the lap, then turn on the drill press. Bring the lap in contact with the brass plate for one or two seconds. The correct amount of pressure is best learned from doing a test piece. I was pleased to find that the amount of pressure applied did not have to be as precisely controlled as I thought it might. It would take an effort to produce a circle much brighter or deeper than the others.

After the first circle is cut, move the sub plate to the left by one index line. Produce the circles in this way across the plate. You will have to fuss a bit with the pieces of the first spacer. The reason, which may now be apparent, is that the lap holder will jam against the first spacer unless it is segmented to allow clearance. At the end of the row, remove the pieces of the spacer. You should now have eight spacers on the base plate. Push the sub plate up against them to establish the spacing of row #2. Move the sub plate back to the far right. You will be using the other register mark for this row. The first circle of each row is positioned so that at least a part of the circle shows on the plate.

Two register marks are used on the base plate so that adjacent rows of circles are not directly over each other. It is essential that you check carefully to make sure that you alternate the register marks with each row. It is easy to check yourself as you go along; just make sure that the circle you are about to make is not directly over the one in the previous row. Figure 103 shows the pattern made by the rows of circles.

At the completion of row #9, the last spacer is removed and row #10 is damascened with the sub plate up against the rear retaining strip. The sub plate is now turned completely around and then moved to the far left to find the beginning of row #11. I found that the new row would have been slightly too close to row #10. This was easily corrected with a thin spacer. To determine its thickness, I installed a pointed center in the drill press. This enabled me to touch the point almost to the plate and see exactly where row #11 would be placed. I tried several thin spacers with the sub plate until I found one that would bring #11 just 5/16" away from #10.

Before you start to damascene the top half of the plate, determine which of the register marks will give you the correct placement of circles. Then begin alternating the register marks with each row as before. Row #11 is done with the thin spacer, if required, then a 5/16" spacer is added before row #12, and so on. For this half of the plate, you will be moving the sub plate one space to the right after

Fig. 103. The damascened effect on the back plate

making each circle. A row of partial circles will be the last one at the top of the plate. Remove the movement plate from the sub plate, and clean the compound off the brass by dabbing with a cloth dipped in mineral spirits. Finish with soap and hot water. Do not touch the plate after this point except with soft gloves or a cloth.

The completed plate should be lacquered immediately to protect the plate from staining and minor scratching. Use either a spray or wiped-on lacquer. I have had good results from wiping on band instrument lacquer. This type dries slower than some lacquers, giving you more time to apply it evenly.

To prepare for this polishing project, I studied several Herschede back plates. The rows appear to have been done in pairs, with the circles interlocking in a way that would be difficult to reproduce. I decided, therefore, to polish one row at a time. There are differences in the spacing of rows on different Herschede clocks. This suggests that Herschede craftsmen did the work in an individual way and that there is no single correct way to do it. It takes about an hour to do each plate; therefore, it is necessary to check the drill press motor from time to time to make sure it is not overheating, and to take several breaks. Interruptions do not spoil the damascening. Try not to touch the circles after they have been polished, as the compound can smear the pattern if rubbed firmly. A light touch will not cause damage. Some of the circles will look more distinct than others, especially after more compound is added to the lap (about every row). However, the circles should appear uniform after the plate has been cleaned.

Fig. 104. A Herschede main
wheel before polishing (below).
The inset shows the same
wheel after polishing.

POLISHING THE MAIN WHEELS

Wheels may need to be polished if they have been
cleaned improperly or allowed to corrode. Most arbors can be gripped in the lathe chuck and the
wheels polished with emery paper. The main wheel
assemblies, however, present a special problem. If
they have been dunked in cleaning solution without being disassembled, the wheels are likely to have
suffered damage from trapped solvents. Figure 104
shows a main wheel that suffered this fate. Hand
polishing did not improve its appearance very much.
Lathe polishing was needed to make the wheel look
as shown in Figure 104 (inset). Four wheels may
require this kind of polishing: they are the strike,
chime, and time main wheels, and the maintaining
wheel.

Two methods were tried for holding the wheels.
One utilized a WW grinding wheel arbor, a threaded
spindle fitted to a collet. I made a brass bushing
which fits over the spindle and allows the wheel to
be mounted. This is shown in Figure 105, on the
left side of the photo. It is certainly useful to have
such an arbor.

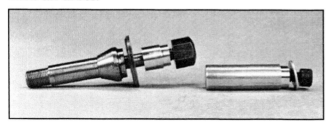

Fig. 105. Two methods for holding a main wheel
in the lathe for polishing: a WW grinding wheel
arbor, fitted with a bushing (left); and a polishing
arbor made from brass rod (right).

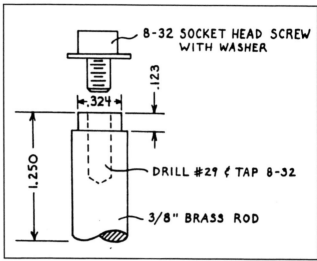

Fig. 106. The polishing arbor for Herschede main wheels
is also seen in Fig. 105 (right side of photo).

It proved just as easy to make a polishing arbor
from brass rod, shown in Figure 106. This arbor
can be held in a 3-jaw chuck in a miniature lathe
such as the Unimat. I have tried the arbor with
equally good fit on Herschede main wheels made in
the 1920's and 1980's. Smooth action is required
for the maintaining power mechanism to operate
on the time main wheel assembly. Beyond this, all
the main wheel assemblies wind much more easily

Fig. 107. Wheel mounted and polished on the Unimat

after being disassembled, cleaned, polished, cleaned
again, and greased. The polishing step is usually
only required if the assembly has suffered improper
cleaning in the past.

It is necessary, of course, to remove the click and
click spring from a wheel before polishing it. Newer
style Herschede main wheels have a brass position-
ing pin for the click spring. This projection makes
polishing more difficult.

16

THE CLOCKS

A number of the Herschede tubular bell clocks from the early 20th century are too tall to even fit in today's homes. This chapter features a later group of clocks more likely to be encountered by collectors and repairers. Each clock was fitted with a Herschede movement of five or nine tubes.

The images of these clocks are reproduced with Herschede's permission, from sales literature of the 1970's and '80's and an older, undated brochure from the 1940's or '50's. Except for the top-of-the-line No. 250 The Clock (right), the Herschede clocks are presented in numerical model order.

Next to each illustration is a catalog description of the clock. Case dimensions, cabinet finishes and woods, and movement choices are provided for each clock.

Fig. 108

250 The Clock

Considered the top of the Herschede product line of fine floor clocks, "The Clock" is given loving care from beginning to completion.

The cabinet is superbly crafted of solids and veneers of olive-ash burl inside and outside. The Old World design is in a class by itself right down to the many feet of antiqued brass trim, which is custom hand fitted. Beveled glass front door and sides, with handsome shell motif and finials atop the pediment. The cabinet is finished in a warm brown color.

The dial is exquisitely hand engraved, hand pierced and gold plated. Note, the dial engraving and piercing is different than other Herschede engraved dials. Raised black numerals, silvered dial ring, with black hands.

The movement is American made by Herschede, 8-day cable wound, three chime melodies played on one and a half inch diameter tubes exclusive in "The Clock". A switch on the dial allows a change from Westminster, Whittington, or *Canterbury chime, also featuring a silent shut-off switch for both chime and strike. Chime tubes available nickel plated or brass, your choice. Pendulum bob and weight shell are brass and lacquered. The brass lyre is a standard feature on the pendulum rod, an option on all other models.

Height 87", Width 23", Depth 15"

*Only on a Herschede Clock.

A handsome Grandfather clock of Italian design, named for the intrepid Italian who discovered America. Cabinet of cherry solids and veneers and beautiful fluted columns. Beveled glass front door and sides.

The dial is hand engraved, pierced and gold plated. Raised black numerals against a silvered dial circle. Dial also features a small second hand.

The movement is made by Herschede and has the traditional moon dial. Chimes are played on tubular bells every fifteen minutes plus counting out the hours each hour in a single strike tubular bell. This movement features three melodies, Westminster, Whittington, and *Canterbury. A switch on the dial allows you to change the melody, also featuring a switch to silence the chimes and strikes if desired.

Pendulum bob and weight shells are of quality brass and lacquered to preserve appearance. A choice of nickel plated or brass tubes available, matter of preference.

Height 86-1/2", Width 24", Depth 15-1/8"

*Only available on a Herschede made clock.

Fig. 109

106 Christopher Columbus

117 Marquis de Lafayette

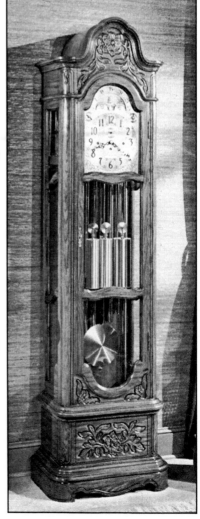

Fig. 110

This fine Grandfather clock exemplifies the French design and social influence in our country. The cabinet is made from solids and veneers of oak with wood grain clearly visible through the fine finish. This cabinet design does not lend itself to beveled glass; therefore, plain is used on front and sides. Top and lower floral designs are machine carved wood panel inserts and not molded plastic as found on lower priced cabinets. The dial is hand engraved, pierced by hand with gold plated finish. Raised black numerals for easy seeing on a silvered dial circle.

The movement is the renowned Herschede eight-day cable wound American made with a moving moon dial; quarter-hour triple chimes, Westminster, Whittington, and *Canterbury. Your choice of nickel-chrome plated tubes or brass tubes. Pendulum bob and weight shells are of brass metal and lacquered to prevent tarnish, satin finished.

Height 86-1/4", Width 24-1/4", Depth 16"

*Only available on a Herschede made clock

Fig. 111

120 Alexander Hamilton

A stately tall-case clock from America's Federal period. Our first traceable effort toward an American national style. Solid and veneers mahogany cabinet. Nine tube, triple chime movement. Moving moon, gold plated, hand engraved dial. 86-3/4" H, 25-3/4" W, 16-1/4" D.

Fig. 112

124 Thomas Jefferson

Classic case design honors one of America's greatest statesmen. Shell carving graces ash wood solids and veneers. Leaded glass front door with amber corner inserts, side glass beveled. Nine tube, triple chime movement. 85" H, 23" W, 17-1/2" D.

Honduras mahogany case with matched sunburst veneer, beaded columns with carved leaf motif at base. 79-3/4" high, 21-3/8" wide, 14" deep. Brown mahogany finish. Nine tube and five tube models.

215 Clairborne

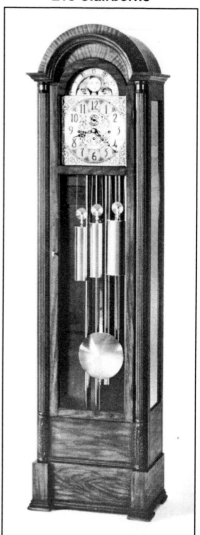

Fig. 113

Fig. 114

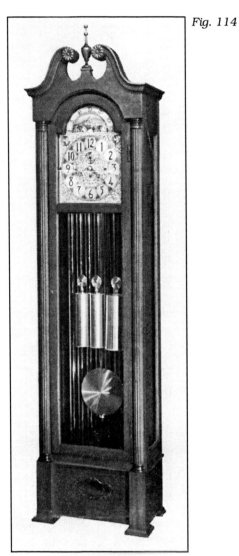

217 Whittier

Delicately beaded columns and overlay trim, all of Honduras mahogany. 80" high, 20-1/2" wide, 12" deep. Brown mahogany finish. Three chimes (9 tubular bells) or single chime (5 tubular bells).

Fig. 115

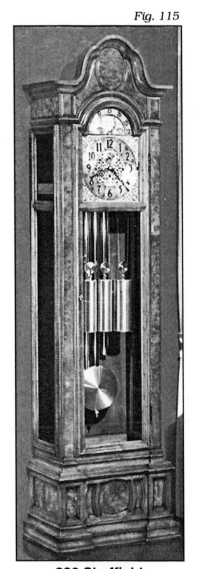

230 Sheffield

The elegance of the late 18th century interpreted in a George I design of solid cherry. Moving moon dial. With 9 tubular bells, triple chimes. Or 5 tube, single chime. 86" H, 24-1/2" W, 15" deep.

Colonial design with full fluted columns. Honduras mahogany or cherry wood and finish. 86" high, 20" wide, 13-1/2" deep. Red or brown mahogany finish available. Westminster chimes (5 tubular bells).

245 Suffolk

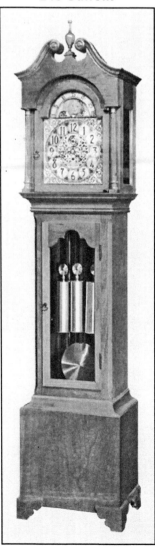

Fig. 116

Fig. 117

247 Edinburgh

A dignified grandfather clock with leaded glass door and sides. Pediment has kingly finials. Solid and veneers mahogany cabinet with book-matched mahogany overlays. 9 tube, triple chime movement. Moving moon, hand-engraved, gold-plated dial. 86" H, 25" W, 16" D.

Fig. 118

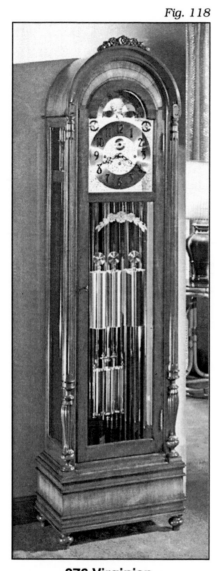

276 Virginian

Avodire (French-named wood) overlays complement mahogany solids and veneers. Floral carving on top accentuates Georgian styling. Acid-etched glass door reveals three-vial pendulum. 9-tube, triple chime movement. Hand-engraved, silver plated dial. Beveled glass sides and main door panel. 79" H, 22" W, 13-1/2" D.

A courtly design reminiscent of the English regency. Solid and veneers mahogany case with book-matched crotch mahogany overlays. Authentic carvings and beveled glass. Moving moon, gold-plated, hand-engraved dial. 9 tube, triple chimes. (Also fruitwood finished cherry) 87"H, 24-1/4"W, 14-3/4"D.

294 Haverford

Fig. 119

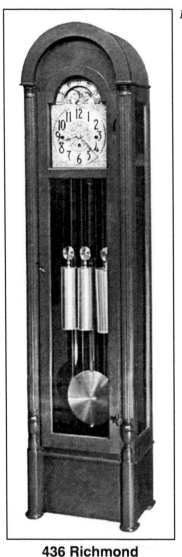

Fig. 120

436 Richmond

A smaller clock with case and beaded columns of Honduras mahogany.
69-3/4" high, 16-1/2" wide, 12" deep. Brown mahogany finish. Westminster chimes (5 tubular bells).

Fig. 121

505 Marietta

Special dimensions make Marietta perfect for condominium and apartment living. Smooth, straight lines with unbroken pediment suggests early Southern influence. 5 tube, single chime. Moving moon, gold-plated dial. Solid cherry and veneers cabinet. 76" H, 20-1/4" W, 13-1/4" D.

A Colonial design with beaded columns and case of Honduras mahogany. 75-1/2" high, 18-3/4" wide, 12" deep. Brown mahogany finish. Westminster chimes (5 tubular bells).

515 Macon

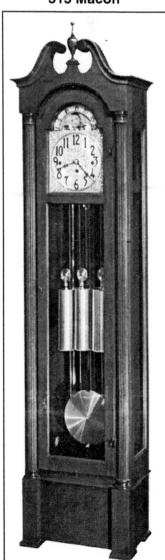

Fig. 122

Fig. 123

525 Antiqua

Old-time richness in high-polished burl veneer finish. Special dimensions for condominium and apartment living. Three glass sides and unbroken pediment. 5 tube, single chime. Moving moon, gold-plated dial.
74" H, 20-1/4" W, 13-1/2" D.

Fig. 124

535 Mount Vernon

Early American tall clock with traditional waisted style. Choice of rich fruitwood finish on cherry wood solids and veneers, or mahogany solids and veneers. Full-view beveled glass on sides and main door panel. Engraved dial, 5 tube, single chime movement.
86-1/4" H, 22" W, 14-7/8" D.

Fig. 125

A touch of English influence handsomely evident in the stately design of the Duke of Marlborough grandfather clock. Cabinet of rich cherry solids with burl cherry overlays with one inch bevel glass sides and door. Exquisite carvings add a touch of elegance to the beautiful Cerisier Antique finish. Left and right movement side access doors with out of sight storage for door key and winding crank. Dial is pierced, hand engraved with silver plated finish. Raised black Arabic numerals and hands, also featuring a small second hand to count off seconds. A choice of three melodies is a feature of Herschede's renown American made nine tubular bell cablewound movement. The Westminster, Whittington and *Canterbury chimes may be changed as desired with striking hour count. Pendulum and weight shells of quality brass and lacquered to preserve the finish. A choice of nickel plated or brass tubes.

Height 88-1/2", Width 28", Depth 17-1/2"

*Only available in a Herschede American made nine tubular bell movement.

8861 Duke of Marlborough

Fig. 126. In 1982, Herschede offered The Earl of Marlborough (below, right) to dealers in a special discount promotion.

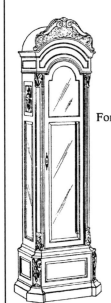

For The Duke of Marlborough with decorative carving order Model No. 8861

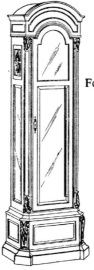

For clock without decorative carving, order The Earl of Marlborough, Model No. 8261

APPENDIX A
HERSCHEDE HISTORY

Frank Herschede was the Cincinnati-born son of German immigrants. He apprenticed in the watch, clock, and jewelry trade and started his own business in 1877, at the age of twenty. By the turn of the 20th century, Herschede had established large, successful clock and jewelry operations.

Herschede's rise coincides with that of another man born the same year. Walter H. Durfee, of Providence, Rhode Island, is often called "the father of the modern grandfather clock". Durfee founded an antique business in 1877, the same year young Frank Herschede started out. By 1884, Durfee had begun to design and assemble ornate hall clocks for the wealthy, reviving an industry that had died out 50 years earlier. Then in 1886 Durfee met the Englishman John Harrington, who had originated and patented tuned sets of nickel plated tubular bells. The tubes produced a resonating, mellow tone described as "the most perfect representation of church chime ever produced." Durfee purchased the rights to act as Harrington's sole U.S. agent for the tubular bells. Harrington, in turn, was a partner in a company which acted as a sales representative for English clock manufacturer J.J. Elliott.[1]

Through these associations, Durfee placed himself in control of a new idea for America: the tubular bell grandfather clock. Herschede's early tubular clocks contained Elliott movements and Harrington tubular bells, apparently all purchased through Durfee. In 1902, Bawo of Brooklyn began to manufacture tubular clock chimes. Durfee sued Bawo for infringement of patent rights, but lost, ending the monopoly. Herschede had been assem-

bling clocks by installing European movements in cabinets ordered from a woodworking shop on Front Street in Cincinnati. Now he was free to develop and manufacture his own tubes, without having to purchase any more material through Durfee or pay license fees.

The business continued to grow, and Herschede "bought out the cabinet shop in 1900."[2] Just after the turn of the 20th century, Herschede entered its hall clocks at the South Carolina and West Indian Exposition and won a Gold Medal.

By 1903, the expanding company had leased factory space at 1011-1015 Plum Street. To put the size of Herschede's clock operation in perspective, however, Richard L. Herschede, Franks's grandson, states in his history of the company that there were eight employees, excluding officers, in the work force early in 1903.[3]

Incorporation

Richard L. Herschede's undated history of the Herschede Hall Clock Company describes the incorporation of the growing Herschede business.

> Mr. Herschede now decided that some more formal organization of the clock activities would be desirable. Before the year (1902) was out, steps were taken to incorporate the business under the name of Herschede Hall Clock Company, "for the purpose of buying and selling, manufacturing and dealing in clocks or parts thereof, and the buying and selling of the raw materials used in the manufacture of clocks or their component parts."

The formal incorporation took place on the 29th of December, 1902.

The steps necessary to complete the organization of the Herschede Hall Clock Company were quickly accomplished in the early part of January, 1903. The five incorporators— Frank Herschede, his younger brother John, and three friends, Edward Greiwe, Francis Pund, and Leo VanLahr— named themselves the first Board of Directors and proceeded with the business at hand. They drew up a simple set of rules and regulations for the operation of the company.

They elected a slate of officers: Frank Herschede, president; Francis Pund, vice-president; and John Herschede, secretary and treasurer.

Initial capitalization of the enterprise was established at $150,000, represented by 1000 shares of common and 500 shares of preferred stock. To provide $15,000 of initial working capital, the directors subscribed to 150 shares of the preferred stock with Frank Herschede taking all but a few token shares.

John Herschede was also named general manager of the company, to have "general supervision of the manufacture and sale of the product, and shall have authority to hire and discharge all employees and to fix their compensation."

The directors also ordered that "the Board shall declare and order paid out of net profits quarterly dividends on the preferred stock at the rate of 6% per annum and such dividends shall be cumulative."[3]

Frank Herschede sold all of his machinery, lumber, clock movements, and "all existing contracts I now have with the manufacturers of Clock movements, and all my agency rights and any and every sort of property I possess connected with the manufacture and sale of Hall clocks and their component parts" to the corporation for $75,000. He accepted company stock as payment, leaving himself in firm control.

The earlier success at the South Carolina and West Indian Exposition was followed in 1904 by another set of awards. At the Louisiana Purchase Exposition in St. Louis, the company won a Gold Medal for the Best Hall Clock, another for the Best Hall Clock Cases, and a Silver Medal for Tubular Chimes.

Richard L. Herschede's history of the company describes how well Herschede continued to do during this period.

At the annual meeting of the company in 1905, the results of the previous year's business were reviewed by John A. Herschede, the secretary, with special emphasis on the company's participation in the St. Louis Exposition. The event was noteworthy not only for the three awards given to the company, but also the publicity and the additional exposure to both the public and the trade resulted in a substantial increase in sales. The general manager was given an increase in salary of nearly 25% and the quarterly dividend of the preferred stock ordered paid through the rest of the year.

In addition to his ownership of the Herschede Hall Clock Company, Frank Herschede was equally successful with his jewelry activities under the name of Frank Herschede Company. By 1905 it had moved into spacious quarters at 24 East Fourth Street, and had become one of the city's leading jewelry establishments. Its advertisements listed "Watches, Diamonds, Jewelry, Umbrellas, Fancy Goods, Bronzes, and articles of vertu."[3]

Manufacturing Herschede Movements

Frank Herschede's son, Walter, was a member of the Board of directors of the company and was appointed treasurer in 1908. He succeeded in obtaining approval to develop the capability for manufacturing movements. In 1909, Herschede leased the building next door at 1007-1009 Plum Street and began to set up for movement production. By the end of 1910, one hundred movements had been completed. The first movement was a tubular model called A-9.[3]

In 1912, Herschede purchased the assets of the Derry Manufacturing Company, of Derry, New Hampshire. Derry's output had included two banjo models, a lyre clock, and an office model.[4] The machinery was shipped to Cincinnati, where it added to Herschede's capacity.

By 1913, Herschede had constructed a new

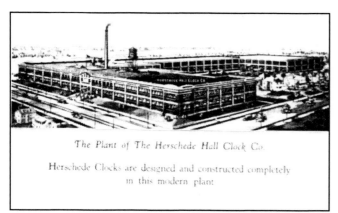

Fig. 127. A view of the factory at McMillan Street and Essex Place, Cincinnati. From a Herschede catalog. Reproduced with permission of Herschede.

manufacturing plant at McMillan Street and Essex Place in Cincinnati (Figure 127). This was the company's main factory for almost 50 years, until the move to Starkville, Mississippi. At about this time, according to Richard L. Herschede's history of the company, a Cincinnati musician named Charles Eisen was commissioned to write a melody to be called Canterbury Chimes.[3] It was included in Herschede triple chime clocks until the end of production in the 1980's.

The Panama-Pacific International Exposition (1915) celebrated the completion of the Panama Canal. Herschede won the Grand Prize for Chime Hall Clocks and Mantel Clocks and a Gold Medal for Hall Clock Cabinets.

World War I brought contracts for surgical and hospital equipment, compasses and surveying instruments".[5] Movement production continued, since it did not depend on German or other European output affected by the war, and there was no problem obtaining raw materials. Herschede continued to do well. It sold movements to some competitors, who could not obtain German movements, including the Colonial Manufacturing Company, of Zeeland, Michigan.[3]

Frank Herschede died on September 22, 1922. His son, Walter, who had been "in complete charge of the manufacturing facilities since before America's entry into World War I"[3], succeeded him as president of the company. Walter Herschede believed in diversification, and he experimented with various products, including the Cleartone radio and two improvements for automobiles—a dashboard mounted fuel level indicator and a wind deflection device.

The Age of Electric Clocks

It was believed in the 1920's that the production of weight and spring driven clocks would be replaced by the manufacture of clocks running on alternating current. Herschede involved itself in 1926, cautiously at first, by founding a separate firm, the Revere Clock Company. The clocks were fitted with Telechron motors made by the Warren Telechron Company. Throughout the rest of the 1920's, electric clock sales, including the sales of electric Herschede floor clocks, increased.

Herschede also made electric clocks, fitted with Telechron motors, under the names Revere and General Electric. In a 1971 letter quoted in an article in the NAWCC *Bulletin*, Jesse Coleman wrote of a visit he made to the Herschede factory in the 1930's:

> The final operation was the application of the hands & dial; they actually came out as three different clocks—"Revere", "Herschede" and "General Electric". It only depended upon which dial the girl fitted on.[7]

Throughout the 1930's, with the country gripped in the Depression, Herschede struggled to keep its workers busy in the production of electric clocks and traditional mechanical clocks. In addition, Herschede began to manufacture parking meters for the Parkrite Corporation. Meters were also made for another company, Karpark Corporation. Herschede modernized its production facilities during this time. In 1938, Parkrite was acquired by Karpark, and Herschede "secured a financial interest in the surviving company. At that time the parking meter business appeared on the threshold of tremendous expansion".[3]

World War II and the Postwar Period

At the beginning of World War II, Herschede changed direction completely by dropping current production in favor of war products. The clock business was limited to making repair parts.[3] A new company, Panocular Corporation, was started to make prismatic lenses for the military. Herschede was very busy with military contracts until the end of the war.

Fig. 128. A letter from the author's files.

THE HERSCHEDE HALL CLOCK CO.

MAKERS OF HIGH GRADE

HALL AND MANTEL CLOCKS

· McMILLAN AND ESSEX PLACE ·

CINCINNATI 6, OHIO, U.S.A.

CABLE ADDRESS
"HERSCHEDE" CINCINNATI

CODES USED
BENTLEY'S-WESTERN UNION

NEW YORK SALESROOM
37 WEST 47TH ST

ROBERT E. WILKES
MANAGER

April 22, 1949

Mr. Frank Ives, Jeweler
725 Kansas Avenue
Topeka, Kansas

Dear Mr. Ives:

Your letter of April 18th received in which you ask for a service
bulletin on our three train #1 clock movement which has a combination
of three chimes and nine tubular bells. We have no service manual
on Herschede key wind weight action movements whatsoever, neither
do we have a parts catalogue or parts price list. If you need any
parts for a Herschede movement we suggest that you send your order,
along with a sample of the old part, to our authorized service
station, Clock Service Center of 1532 Madison Road, Cincinnati 6,
Ohio, and they will be glad to take care of your order for these
parts.

We manufacture, in addition to Herschede floor clocks, a complete
line of synchronous electric mantel chime and tubular chiming floor
or Grandfather clocks, also Grandmother clocks, which are merchan-
dised through our subsidiary, The Revere Clock Company. There is
a service manual on these electric clocks and for your information,
one of the manuals is attached to this letter.

Going back again to the Herschede movements, if you have one of our
movements which you are repairing and you want some added information, give us the number of the movement which is stamped on the
back movement plate near the bottom, and outline in a letter the
difficulty you are having. We will endeavor to suggest the necessary adjustment to be made via return mail.

Yours very truly,

THE HERSCHEDE HALL CLOCK COMPANY

By

EEK/me
encls.

P.S. Any parts for Revere clocks should be also ordered through Clock Service
Center of the above address.

After World War II, the parking meter business revived, but competition soon reduced the profitability of this work. To make matters worse, General Electric purchased Telechron and dropped Herschede as a supplier of electric clocks, ending a large account spanning many years.

A new Herschede facility called the Clock Service Center was opened in 1948 to handle "repair, adjustment, and service needs". Walter Herschede, Jr. was the manager of this new organization. Figure 128 shows a letter written to a Kansas jeweler the following year, referring to this new facility.[3]

The 1950's

Richard L. Herschede's history of the company describes the changing economic times of the 1950's. European clock factories, including those making clocks, were being rebuilt, and although still protected by tariffs, U.S. manufacturers anticipated the end of these tariffs. They expected stiff price competition from abroad. Walter L. Herschede traveled to Europe in 1950 "to investigate the possibility of securing various parts and movements from sources there." Soon after, the Korean conflict began and the company reactivated the Panocular Corporation to manufacture tank periscopes. A planned addition to the plant in Cincinnati enabled Herschede to continue manufacturing clocks and parking meters and to anticipate more war work.

At the same time, Walter Herschede went to Europe and completed arrangements with Gebruder Junghans of Schramberg, West Germany, to have them supply not only clock movements but complete clocks which would be imported under the Herschede name for sale in the domestic market. He also gathered information on European sources for other metal clock parts, cases and dials, in the event these should be needed to supplement or take the place of Herschede manufactured items.

By early 1952 the new addition to the plant was completed, and production of tank periscopes was proceeding in the new location. In the old plant, production facilities were largely devoted to making Karpark meters. Clock production was limited to hall or floor clock models; others carried in the line were those imported under the Herschede name.[3]

The Move to Starkville

With the end of military production, Herschede's production declined. In 1957 and 1958, the company considered several plant sites where manufacturing costs would be lower than in Cincinnati. In 1959, Starkville, Mississippi was selected as the new site. Richard L. Herschede says in his history that in April 1960 a few supervisory and office employees were given the option to move with the new operation. Walter Herschede, approaching 75 years of age, retired. The dedication of the new plant took place on May 17, 1960, and "complete facilities were provided in the new plant for the manufacture of the company's famous line of tubular chime clocks."[3]

Herschede had purchased Rookwood Pottery Company, and the pottery was made in a section of the new plant. In addition, a military contract was obtained for the making of telescope mounts. Another product was the "Motor-Guide" electric trolling motor for small boats.

Richard L. Herschede became president of the company in 1962.

Beginning of the End?

Despite the efforts to diversify, the company's financial position declined, and in 1967 Herschede became a division of Arnold Industries Inc. Operations continued, but the company was again sold in 1984, this time to Brunswick Corporation, which was interested primarily in the Motor Guide business. Clock production became dormant, although efforts were made to sell Herschede tubular bell movements without cases. For a time, Herschede tubular bell movements were offered to American clock companies, for use in their own cabinets. They were also offered to the trade; a 1985 catalog page from S. La Rose, Inc., a North Carolina clock parts supplier, offered the complete 9-tube movement with dial, hands, weights, tube rack, seatboard, and bells, for $1795.00. There was no denying, however, that 99 years after Frank Herschede assembled his first tubular bell clock, Herschede was no longer in production.

Revivals

In 1988 Howard Klein, a St. Louis clockmaker, and his partner, Robert Eggering, purchased the remaining assets of the Herschede operation, including

Fig. 129. A revived Herschede, under the leadership of Howard Klein, produced three hall clocks. A 100th Anniversary Edition (right) was displayed at the 1990 High Point Market. The other two models are shown below. Photos courtesy of Howard Klein.

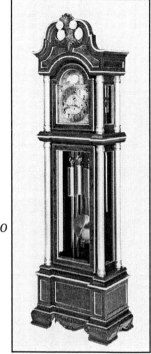

The Herschede 100

246II Edinburgh

294II Haverford

some of the tooling and many parts, drawings, and company records. The new company soon offered Herschede spare parts and a repair service for the tubular bell movements. The owners' dream was to resume the manufacture of the hall clocks. Production was begun on three models (Figure 129), but only a relatively few clocks were produced. The new enterprise was unable to locate a ready market for these "high-end" clocks that were priced well above the rest of the new floor clocks offered at the time. In 1992, Herschede was sold to Randy Thatcher, an Ohio clock parts supplier.

While operating Herschede as a parts business, Randy has manufactured many individual parts that had been unavailable for years. In addition, he has done much to keep Herschede history alive and to research little known aspects of the company's century of clock production.

Randy and his wife have checked on some of the historic Herschede locations in Cincinnati. They found that the site of the Arcade, where Frank Herschede operated his retail business in 1877, is now a parking lot on the west side of Cynergy Field. The Plum Street factory is an office building. Frank Herschede's Cincinnati mansion is still there—with its high ceilings and brocaded walls—and it is in good condition. The main factory at McMillan Street and Essex Place is in the Time Hill district and is now a garment factory. Minus the old water tower, Randy says the factory still looks much the same as in the old illustrations.

Sources

1 NAWCC *Bulletin*, Whole Number 215, December 1981, "Walter H. Durfee, His Clocks, His Chimes, His Story", by Owen and Jo Burt, pp. 556-583.

2 "A Brief History of the Herschede Hall Clock Company", an undated company brochure.

3 An undated, 58-page, typed history of the company written by Richard L. Herschede.

4 NAWCC *Bulletin*, Whole Number 189, August 1977, Answer Box item on the Derry Manufacturing Company, page 383.

5 *150 Years of Electric Horology*, Midwest Electric Horology Group, NAWCC Chapter #125, 1992, edited by Elmer G. Crum and William F. Keller, article "The Herschede Hall Clock Company" by Howard Klein, pages 35-36.

6 NAWCC *Bulletin*, Whole Number 255, August 1988, article "Herschede: From Beginning to End" by John E. Huett, pages 329-330.

7 NAWCC *Bulletin*, Whole Number 265, April 1990, article "Tidbits from J.E. Coleman" by James W. Gibbs, page 133-134.

APPENDIX B

HERSCHEDE TEST STAND PROJECT

A tubular bell movement, with its extra-large weights and thick plates, is too heavy for some test stands to accommodate safely. Making matters worse, the Herschede tubular bell movement has the tubes hanging from a rack attached to the seatboard. The Herschede movement can be tested with the tubes, which is important for a valid test of the chimes and strike. This creates the need for a stand that has the dimensions and strength necessary to handle the whole mechanism, including the tubes and the heavy dial.

This appendix includes the plans for a wooden, factory test stand designed specifically for Herschede tubular bell movements. With help from friends skilled in woodworking, I made the stand shown in Figure 130. The illustrations in this appendix are based on a Herschede blueprint, plus a sketch and photos of my stand. I made two modifications to the factory design, and these will be explained.

As you consider this stand for your own shop, be aware that it is quite bulky, standing over 49" tall, 28" wide, and 20" deep. It is ideal for large, seatboard mounted grandfather clock movements, including tubular bell and English bell strike types. As an added benefit, the top of the stand offers enough area to lay down your tools as you work on adjusting a movement. The sturdy construction makes it unlike any test stand I have seen.

The test stand was constructed of 3/4" plywood and pine lumber in 1 x 4 and 2 x 4 sizes. The joints were screwed together. The legs were given dado cuts to make the joints with the rails stronger, although this is not called for in the blueprint.

The first change I made was to make a third front

rail. Two rails are specified in Figure 132. An extra rail was mounted at floor level, across the back legs, to strengthen them. This modification can be seen in Figure 130.

Fig. 130. Herschede test stand with General Electric (Herschede) electric tubular movement mounted.

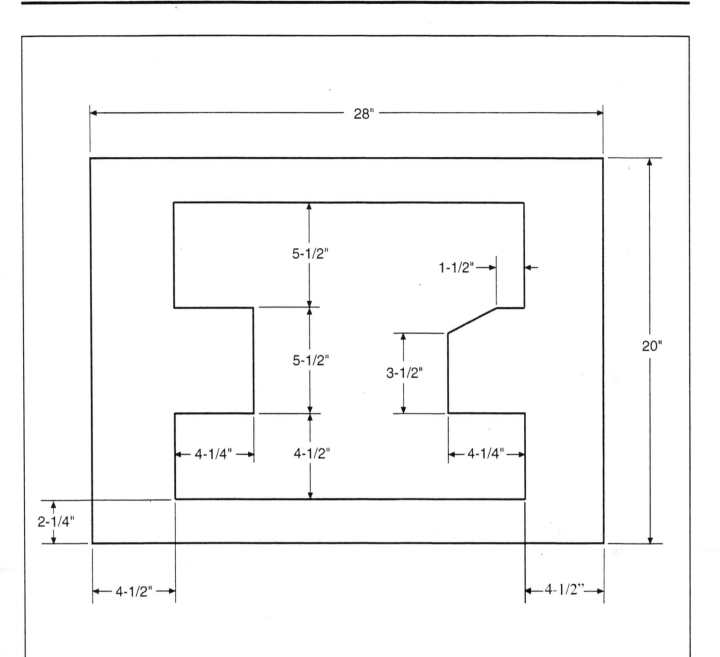

HALL CLOCK TEST STAND

TOP

MAKE ONE
3/4" X 20" X 28" (PLYWOOD)

Fig. 131. Drawing of the top piece. Redrawn from a Herschede blueprint, used with permission.

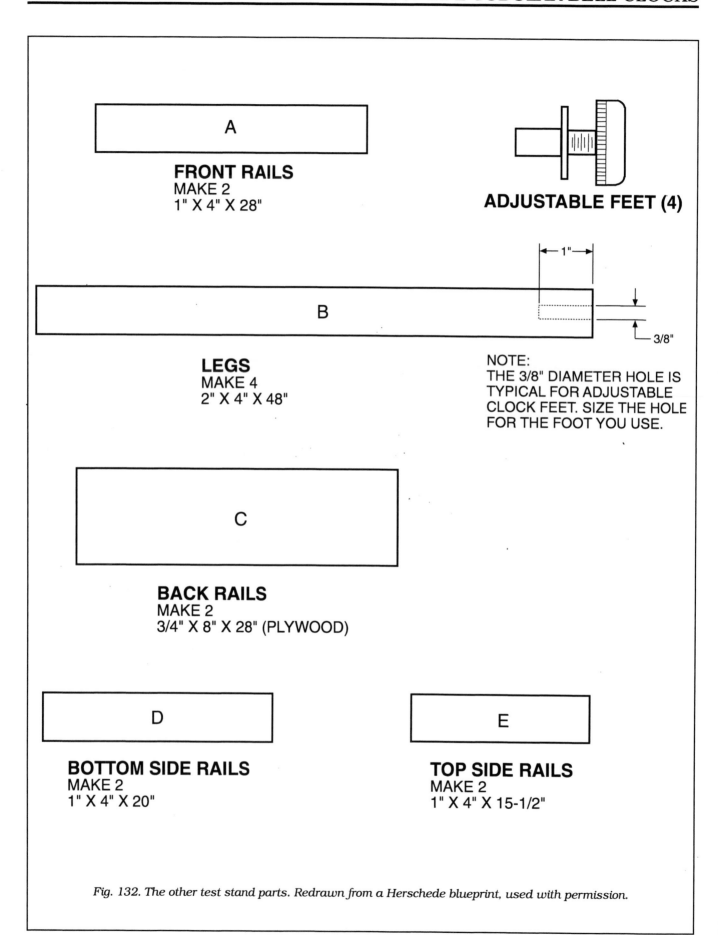

Fig. 132. The other test stand parts. Redrawn from a Herschede blueprint, used with permission.

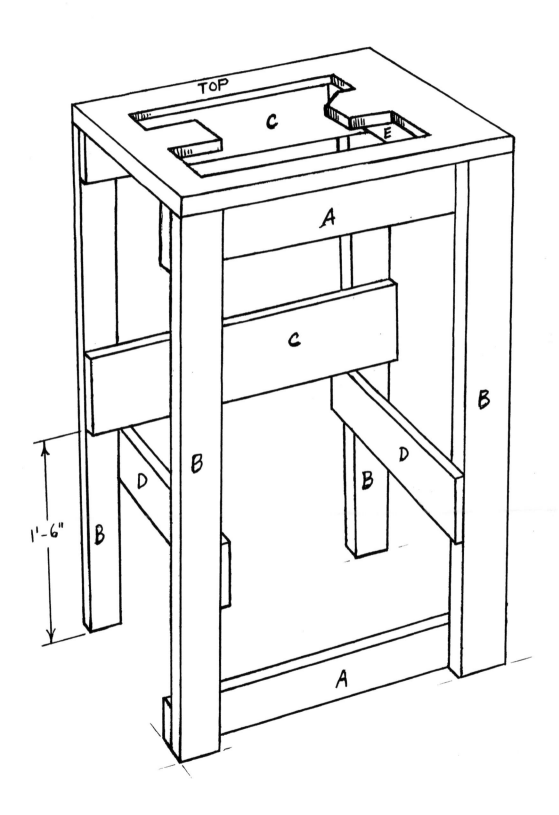

Fig. 133. Sketch of the Herschede test stand. The two modifications explained in the text are not shown on this sketch.

*Fig. 134. This side view shows the location of the top side rails **E**. Just to the right of the upper letter E in the photo can be seen one of the two added vertical braces for the top (see text below).*

The Second Modification

The top piece (Figure 131) has two projections which extend into the central cutout. These projections support the clock's seatboard and yet allow room for the dial, at the front, and the tubes, at the back. The projection on the right is beveled to allow clearance for a Herschede hour tube to hang freely from the tube rack.

The second modification I made to the test stand was to add a pair of vertical braces to the underside of the top piece, to support the projections. The braces were made from rectangular pieces of 1 x 4 wood left over from the project. By screwing them to the test stand top and to the top side rails **E**, I was able to add strength to the projections. This is especially helpful when movements with shorter seatboards are mounted, since additional stress is placed on the projections.

The author thanks Ron Pfleger for helping with the transportation of the wood, the initial layout work, and the cutting of much of the wood. For various reasons involving health, the seasons, and being just plain busy, we left the project uncompleted for some time. When I finally returned to it, Ed Sowers helped out with tools and expertise in cutting the top piece and in getting the pieces joined together. For a novice woodworker, I did pretty well, managing in the end to get the stand together. In the process, I inflicted upon myself the worst splinter I've ever had in my thumb. No doubt some readers will make a faster, less painful job of it. Please use safe work practices in all such projects.

My appreciation also goes to Randy Thatcher of Herschede, for allowing these plans to be published.

APPENDIX C

SERIAL NUMBER LISTING

Herschede serial number lists are incomplete, and it is uncertain whether more detailed information will come to light. Nonetheless, the numbering in this appendix will help to date many Herschede clocks. The serial number is generally found on the movement back plate.

This serial number information has previously been published by Herschede in pamphlet form and is presented here with permission.

Through 1910, Herschede imported European movements.

1911
1-66
101-141
151 -200
214-234
243-246

1912
67-100
142-150
201-213
235-242
247-400

1913
401-536
551-602
1,001-1,491

1914
537-550
603-816
1,492-1,802
1,853-2,302

1915
817-1,000
1,803-1,852
2,303-2,600
2,666-2,670
2,901-3,218

1916
2,601-2,604
2,769-2,792
2,806-2,847
3,219-4,233

1917
2,605-2,664
2,671-2,750
2,793-2,801
2,848-2,900
4,234-5,122
5,232-5,556
7,361-7,445

1918
2,751-2,768
2,802-2,805
5,123-5,231
6,555-7,056
7,338-7,360

1919
7,057-7,337
7,455-7,466
7,476-7,535
7,801-8,800

1920
6,057-6,554
7,467-7,475
7,701-7,800
8,801-10,350
10,459-10,477
10,651-10,760
10,851-11,900
12,401-12,600
12,651-12,800

1921
7,446-7,454
7,536-7,559
10,401-10,425
10,901-11,950
12,351-12,400
12,601-12,650
12,801-12,850
13,001-13,500
15,001-15,117
15,801-16,204

1922
10,363-10,400
10,426-10,450
10,478-10,500
10,761-10,813
12,077-12,100
12,156-12,266
12,851-12,950
13,501-13,732
14,017-14,150
15,118-15,350
16,205-16,669

1923
7,665-7,700
10,351-10,354
10,501-10,534
11,951-12,076
12,101-12,155
12,268-12,350
13,733-14,016
14,151-15,000
15,351-15,800
16,670-17,600
18,001-18,450
18,501-19,000
19,101-20,500
21,701-22,500
24,701-24,850

1924
10,355-10,362
12,951-13,000
17,601-18,000
18,451-18,500
19,001-19,100
20,501-21,780
22,501-23,400
23,501-24,700
24,851-30,000
36,001-36,350

1925
7,560- 7,664
10,535-10,555
10,814-10,850
23,401-23,500
30,001-32,000
32,051-32,100
23,122-32,200
32,299-32,415
32,442-32,486
32,513-32,750
33,001-34,000
35,501-36,000
36,351-38,400

1926
10,451-10,457
32,001-32,009
32,106-32,121
32,201-32,298
32,490-32,503
32,751-33,000
34,002-34,004
34,008
34,021
34,029
34,034-34,035
34,101-34,121
34,257-34,258
34,268-34,272
34,282-34,283
34,286-34,290
34,294-34,297
34,302-34,304
34,313-34,316
34,323-34,324
34,327-34,329
34,363-34,366
34,477-34,480
35,101-35,300

1927
10,566-10,575
32,010-32,029
32,417-32,441
34,013-34,014
34,025-34,027
34,039-34,040
34,045-34,049
34,370-34,372
34,395-34,398
35,001-35,100
35,201-35,500

1928
32,030-32,050
32,504-32,505
79,101-99,900

1929
32,101-32,105
·32,506 -32,512
99,901-123,500

1932
250,001-255,700

1933
255,701-261,000

1934
113,001-215,600
261,001-269,381

1935
271,382-285,268

1936
285,269-298,323

1937
298,725-313,243

1938
313,244-315,745

1939
316,247-325,376

1941
347,103-361,368

1942
362,135-394,133

1945
386,000-407,326

1946
401,077-424,068

1947
424,069-489,768

1948
105,552-130,576
391,000- 506,000

1949
130,577-151,226 ·
512,001-521,535

1950
151,227-163,478
521,536-539,535

1951
163,479-175,658
539,536-558,535

1952
175,659-187,908
558,536-564,535

1953
187,909-194,158
564,536-582-535

1955
589,036-595,035

1956
595,036-603,965

1958
603,966-604,174

1959
604,501-605,000

1961
605,001-605,843

1963
605,895-616,902

1964
612,446-616,882

1965
616,904-617,405

1966
617,403-620,403

1967
620,404-623,403

1968
623,404-625,903

From 1969 to the end of production, no records are available; however, movements with serial numbers preceded with the letter "A" were made after the 1980 retooling.

INDEX

ACKNOWLEDGMENTS

The 1986 first edition paid tribute to several people:

Jim Craven, who has since passed away, was my first clock repair teacher.

Al Stevens has been a source of information and advice over the years.

The magazines *Watch & Clock Review* and *Horological Times* published early versions of some of the chapters during the early 1980's.

My wife, Karen, whom I thanked for her encouragement during the writing of the first edition, was also a great help with the second. She reads each of my manuscripts and newsletters before publication, helping to find the kinds of errors that the author usually can't see after so many hours spent on a project.

To this list I would add two others who have helped with the second edition:

Howard Klein is the former Herschede owner who worked tirelessly to restart the manufacturing of Herschede hall clocks. Howard has provided much helpful information, especially for the years during which he headed the company.

Randy Thatcher, the present-day owner of Herschede, has been supportive in every way. Randy has given permission for the use and reproduction of any company material in his possession. His support of the second edition has been essential. In particular, his reading of the manuscript identified areas for correction and greatly improved the resulting book.